How Things Are

How Things Are

A Science Tool-Kit for the Mind

John Brockman and Katinka Matson, Editors

Quill   William Morrow   New York

Library of Congress Cataloging-in-Publication Data

How things are : a science tool-kit for the mind / John Brockman
    and Katinka Matson, editors. — 1st ed.
        p.   cm.
    ISBN 0-688-14951-0
    1. Science—Miscellanea.   2. Science—Philosophy—
    Miscellanea.   3. Science—Methodology—Miscellanea.
    4. Evolution—Miscellanea.   5. Mind—Miscellanea.
    I. Brockman, John, 1941–   II. Matson, Katinka.
    Q173.H742   1995
    500—dc20   94-40003
                                                                  CIP

Printed in the United States of America

First Quill Edition

1 2 3 4 5 6 7 8 9 10

BOOK DESIGN BY CHRIS WELCH

For our son, Max

Contents

Introduction

How Things Are is a gathering together of some of the most sophisticated and distinguished scientists and thinkers in the world, each contributing an original piece—an elementary idea, a basic concept, a tool for thought—relevant to his or her field of science.

The appeal of the essays is their focus and their brevity: a single theme; a surprising point of view; an explanation about why a theory is acceptable or not.

How Things Are gives us a chance to see the minds of distinguished scientists and thinkers at work: the questions they ask, the methods they use; the thought processes they follow in trying to arrive at an understanding of the world around us, and of ourselves.

Imagine that you found yourself in a roomful of distinguished scientists, and that you were able to pose just a single question to each one. These essays can be read as their responses to your questions—written in a language that you can understand. What a rich resource, this gathering of important thinkers.

Some of the questions whose answers we find in the book are: What's natural? Why are there differences in skin color? What are the biological bases of racial differences? How do we understand the current world population, through history and into the future? What is the difference between mind and brain? What's good about making mistakes? What are the differences between living and nonliving matter? What is evolution? Why incest? How is human communication possible? How do computers change the way we think? When did time begin?

Reading the essays is like overhearing a conversation among scientists dining at a nearby table. You will discover not only what topics get their attention, but also how they formulate the questions they seek to answer. Most important, you will come away from the book with a deep sense of how scientists inquire and how they discover *How Things Are.*

In "Part One: Thinking About Science," we find zoologist Marian Stamp Dawkins on scientific explanation; anthropologist Mary Catherine Bateson on the concept of "natural"; and evolutionary biologist Richard Dawkins on believing.

"Part Two: Origins," features theoretical physicist Paul Davies on the "big bang"; chemist Peter Atkins on water; chemist Robert Shapiro on the origins of life; biologist Jack Cohen on DNA; biologist Lewis Wolpert on the miracle of cells; biologist Lynn Margulis on what a drink called kefir can teach us about death.

In "Part Three: Evolution," zoologist Stephen Jay Gould writes on the idea of evolution; anthropologist Milford Wolpoff on the relationship between humans and apes; anthropologist Patrick Bateson on the development of the incest taboo; biologist Steve Jones on why people have different skin colors; paleontologist Peter Ward on the role chance plays in evolution; and evolutionary biologist Anne Fausto-Sterling on the normal and the natural versus the abnormal and the unnatural.

In "Part Four: Mind," philosopher Daniel Dennett reports on the importance of making mistakes; logician Hao Wang on the mind versus the brain; neurophysiologist William Calvin on ways to think about thinking; neurobiologist Michael Gazzaniga on the unique patterns of each brain; anthropologist Pascal Boyer on thinking tools; psychologist Nicholas Humphrey on illusion; psychologist and artificial-intelligence researcher Roger Schank on learning by doing; anthropologist Dan Sperber on human communication; neuroscientist Steven Rose on the mind/brain dichotomy; computer scientist David Gelernter on learning how to read; and psychologist Sherry Turkle on how computers change the way we think.

In "Part Five: Cosmos," we find physicist Lee Smolin on the question of time; physicist Alan Guth on the importance in science of thinking the impossible; mathematician Ian Stewart on the symmetric patterns of nature; and computer scientist Daniel Hillis on why we can't travel faster than light.

In "Part Six: The Future," Freeman Dyson writes on the problem of scientific predictions; population mathematician Joel Cohen on why Earth's present population growth is unique; paleontologist Niles Eldredge on why the world is *not* going to hell in a handbasket; and cosmologist Martin Rees on understanding the stars.

JOHN BROCKMAN and KATINKA MATSON
Bethlehem, Connecticut

Part One

Thinking About Science

Nothing But or Anything But?

Marian Stamp Dawkins

Are you one of those people who dislikes science because of what you think it does to the "mystery" of the world? Do you think that it somehow diminishes people to try to "explain" their behavior or that it "takes away the beauty of the rainbow" to talk about light rays passing through water droplets?

If you are, can I ask you to pause just for a moment and very briefly allow your mind to entertain the opposite view, perhaps one that will strike you as utterly preposterous and even repugnant? You can, if you like, think of yourself as being like the captured prince in C. S. Lewis's story *The Silver Chair*—the one who had to be tied down every evening at six o'clock because, for an unpleasant few minutes, he had hallucinations and then recovered and was perfectly normal for the next twenty-four hours. The hallucinations will not last long, in other words. You are quite safe. Just five minutes of raging lunacy is all you are in for.

The utterly preposterous idea is this: Explaining something in a scientific way does not diminish it. It enhances it. Let me tell you why. Understanding how things work, even your own brain, has a grandeur and a glory that no nonscientific explanation can come anywhere near.

I do not, of course, expect you to accept this without question. But I do ask that you start by thinking of something reasonable and not scientific at all—like, say, Abraham Lincoln. Ask yourself whether it enhances or diminishes your view of his achievement when you remember that he was "nothing but" a country-born, self-taught lawyer. Then ask yourself which you admire most: someone who starts from an unpromising background and achieves great things by his own efforts and personality, or someone who comes from a wealthy and powerful family and achieves high office because he has an influential father. I would be surprised if Abraham Lincoln didn't come out of that comparison very well. Saying he is "nothing but" a backwoodsman doesn't get rid of him that easily.

Next, try the pyramids. Does it diminish the achievements of the ancient Egyptians to say that they built them with "nothing but" the crudest of tools and measuring instruments? My own sense of awe and admiration for them only went up when I realized that they moved gigantic blocks of stone with no wheeled vehicles and that they built the pyramids with perfectly square bases using "nothing but" lengths of knotted cord and stakes in the ground. Even the slightest error would have led to the whole structure being hopelessly out of shape, and yet there they remain to this day—phenomenal feats of engineering and nearly perfect shapes. "Nothing but" simple equipment becomes "anything but" a mean achievement.

Perhaps by now you can see what I am getting at. If you

look at a rainbow and then someone tells you how it comes about, why should you say that "they have ruined it" by turning it into "nothing but" a trick of light and water? Why not say the opposite? Out of the utterly unpromising raw material of drops of water and the laws of refraction has come something so beautiful that it spans the sky and inspires poets to write about it.

And if you look at animals and plants and all the extraordinary structures and behavior they have, why reject a scientific explanation of how they got here on the grounds that it makes them "nothing but" the products of a blind evolutionary process? You could instead turn and face the full grandeur of the implications of what that process implies. The bird that builds a nest and brings food to its young may be "nothing but" the result of evolution by natural selection. But what a result! Birds, just like us, owe their existence to instructions carried on DNA molecules. The scientists in *Jurassic Park* who grew dinosaurs out of DNA molecules that were found in preserved blood were on the right track, even though nobody has yet done this in practice. DNA molecules do indeed carry the instructions for building all kinds of bodies—dinosaurs, birds, giant sequoias, and even human beings.

And the molecules do not stop there. Every breath you take, every extra second you remain alive depends on hundreds of chemical reactions all taking place at the right time. Life would be impossible, for instance, if the body did not have a constant source of energy—and that energy comes from molecules such as glucose. Glucose can provide energy, because the three different sorts of atoms that make it up—carbon, hydrogen, and oxygen—are kept together by energy-rich links between the atoms, called chemical bonds. If some of these bonds are then broken so that the large glucose molecule is split into smaller molecules of water and carbon dioxide, the energy that was once used to hold the whole glucose molecule together is then released. So the body plunders the molecules by breaking them up and steals their energy to keep itself going. It could do nothing unless it was constantly fueled by this energy. It would grind to a halt, lifeless and inert. You could not move a muscle or think a thought without this constant molecular smashing that your body does and without the energy from the broken molecules being transported by yet other molecules to the places where it is needed. The bird could not build its nest and we could not watch it or wonder why it did so. Molecules have gone far, considering that they are "nothing but" molecules.

There! The few minutes of hallucination are up. I'll untie you now and you can go back to thinking that scientific explanations diminish and belittle everything they touch. There is just one thing, though. The prince in C. S. Lewis's story had rather an unusual sort of hallucination. For those few minutes each day, he suffered from the vivid delusion that there was something beyond the dark underground kingdom in which he had been living. He really believed— poor demented soul—that there was something called sunlight and a place where the sky was blue and there could be a fresh breeze on his cheek. But that was only for a few minutes. And then it passed.

MARIAN STAMP DAWKINS has a lifelong interest in what the worlds of different animals are like—that is, not just in what they can see, hear, or smell, but in what they know about their worlds and, above all, in whether they are conscious of what they are doing. At the same time, she firmly believes that answers to these issues should come not from anthropomorphism but from scientific research and, in particular, from studies of animal behavior.

Much of her own research has been concerned with animal welfare and, in particular, with the problem of whether animals can experience "suffering." She has published extensively in this field, including the book *Animal Suffering: The Science of Animal Welfare.* She has also written *Unravelling Animal Behaviour* and *An Introduction to Animal Behaviour* (with Aubrey Manning). Most recently, she published *Through Our Eyes Only? The Search for Animal Consciousness.* Her current research concerns the evolution of animal signals, how birds recognize each other as individuals, and why the fish on coral reefs are so brightly colored. She holds a university research lectureship in the Department of Zoology at Oxford University and is a fellow of Somerville College.

On the Naturalness of Things

Mary Catherine Bateson

~~~~~~~~~~ Clear thinking about the world we live in is hindered by some very basic muddles, new and old, in the ordinary uses of the words *nature* and *natural*.

We seem to slip easily into thinking that it is possible to be outside of nature—that with a little help from on high, we could rise above the ordinary contingencies, evade the consequences of our actions, and be supernaturally delivered from the all-too-natural realities of illness and death. Some usages seem to suggest that it is possible to be below nature, as in "unnatural acts" (sometimes called "subhuman"), or "unnatural parent" (which means an unloving parent or one who fails in the obligations of nurture, with no logical connection to a "natural child," one born outside of the culturally sanctioned arrangement of wedlock).

These usages have in common the notion that nature is something it is possible to get away from, to get around. The intellectual problems created by circumscribing the domain of "nature" are probably even more confusing than those created by Cartesian dualism, although they are no doubt related. Descartes was concerned to define a domain for science that would be safe from ecclesiastical interference: *res extensa,* matter, the physical body, divorced from mind or spirit. The effect of this was to create two different kinds of causality and separate spheres of discourse that must someday be brought back together. The folk distinctions that describe the concept of "nature" are messier but equally insidious. As with Cartesian dualism, they tend to slant ethical thinking, to create separation rather than inclusion. In Western culture, nature was once something to be ruled by humankind, as the body was to be ruled by the mind.

Recently, we have complicated the situation by labeling more and more objects and materials preferentially, from foodstuffs to fibers to molecules, as natural or unnatural. This sets up a limited and misshapen domain for the natural, loaded with unstated value judgments: the domain suggested in Bill McKibben's title *The End of Nature* or in William Irwin Thompson's *The American Replacement of Nature.* Yet nature is not something that can end or be replaced, anymore than it is possible to get outside of it.

In fact, everything is natural; if it weren't, it wouldn't be. That's *How Things Are:* natural. And interrelated in ways that can (sometimes) be studied to produce those big generalizations we call "laws of nature" and the thousands of small interlocking generalizations that make up science. Somewhere in this confusion there are matters of the greatest importance, matters that need to be clarified so that it is possible to argue (in ways that are not internally contradictory) for the preservation of "nature," for respect for the "natural world," for education in the "natural sciences," and for better scientific understanding of the origins and effects of human actions. But note that the nature in "laws of nature" is not the same as the nature in "natural law," which refers to a system of theological and philosophical inquiry that tends to label the common sense of Western Christendom as "natural."

Ours is a species among others, part of nature, with recognizable relatives and predecessors, shaped by natural selection to a distinctive pattern of adaptation that depends on the survival advantages of flexibility and extensive learning. Over millennia, our ancestors developed the opposable thumbs that support our cleverness with tools, but rudimentary tools have been observed in other primates; neither tools nor the effects of tools are "unnatural." Human beings communicate with each other, passing on the results of their explorations more elaborately than any other species. Theorists sometimes argue that human language is qualitatively or absolutely different from the systems of communication of other species; but this does not make language (or even the possibilities of error and falsehood that language amplifies) "unnatural." Language is made possible by the physical structures of the human nervous system, which also allow us to construct mental images of the world. So do the perceptual systems of bats and frogs and rattlesnakes, each somewhat different, to fit different adaptive needs.

It is often possible to discover the meaning of a term by seeking its antonym. Nature is often opposed to culture or to nurture. Yet human beings, combining large heads and the appropriate bone structure for upright posture and bipedal locomotion, have evolved to require a long period of adult care after birth, time to acquire those variable patterns of adaptation and communication we call culture. How then could "nurture" be "unnatural"? The characteristics of the human species that set us at odds with our environment and with other species are part of the same larger pattern.

Increasingly, nature is opposed to artifact, yet, human beings must always work within natural possibility to create their artifacts, even in the productions of dream and fantasy. Ironically, in current parlance, many artifacts are called "natural." If what we mean by "natural" is "unaffected by human acts," the natural is very hard to find. Walk in the woods, for instance. Patterns of vegetation in different North American biomes were already changed by human habitation long before the first Europeans arrived, and were changed again by the colonists. Today, there are introduced species of birds and insects and plants all across this country, even in so-called wilderness areas. The migrations of human beings to every continent on the globe have transported human parasites and symbionts since prehistoric times. Human beings, as they learned to use fire, weapons, and agriculture, have exerted selective pressures everywhere they have lived—like every other species. Henry David Thoreau was fully aware that what he could study and reflect on, living beside Walden Pond, already bore a human imprint. Still, we are wise to treasure and learn from landscapes in which the human imprint is not obvious. This is perhaps what we usually mean by wilderness (one wonders how much the wilderness into which Jesus or John the Baptist withdrew was a human creation, as so many spreading deserts are today). Wilderness turns out to be a relative term but still a valuable one. We need areas with no visible structures and no soft-drink cans to remind us of human activity, but still, they are affected by human acts.

If natural means "unaffected by human acts," it won't be found at the "natural foods" store. Most food products have been produced by selective breeding over the centuries, turning wild plants into cultivars dependent on human beings and multiplying or eliminating their variations. Most are also processed and transported in clever cultural ways; after all, tofu doesn't grow on trees. Organic farmers must work hard and skillfully; nature doesn't do their work for them. Still, the effort to produce foodstuffs without the chemical fertilizers and insecticides that produce toxic residues is an important area of ingenuity and persuasion. It would be nice to find a way of talking about it without nonsensical and self-contradictory uses of terms like "natural" or "organic" (what would a vegetable be if it wasn't "organic"?). Some of the animals and plants cultivated by human beings can survive without human help, like domestic house cats that become feral, foreign to their settings, and disruptive to other species. Living in a more "natural" way, they may be more disruptive. It may be useful to distinguish between what we create "on purpose" and unexpected by-products. In this sense, gardens of any kind should be distinguished from the deserts created by some of the ways in which humans use land.

Human populations today exist because of massive interference with "nature." Without the invention of agriculture and other technologies, human populations would have stayed tiny, and most of our ancestors would never have been born.

Individually, we are probably alive because of medical technologies, public health, and immunizations. Without some kinds of technology, you're dead. When warfare disrupts the artifices of public health, clean water, transport, electricity, and so on, the death rates reflect this new level of "naturalness." Even "natural childbirth" is an invention that depends on modern ideas of hygiene, training, emergency backup—and on the use of a watch to time contractions. Some of us pride ourselves on looking and acting "natural," but try looking in the mirror. Do you use hair conditioner, toothpaste, vitamins? Even the so-called natural ones are human artifacts—and so are a clear complexion, shining hair, straight teeth.

All this will become clearer if we try looking at something really "unnatural": a hydroelectric dam, for instance, or a plastic bag, a nuclear plant, or a polyester suit. All of our artifacts exist only because they fit into natural possibility— sometimes all too well. If they did not, they would not serve our purposes; bridges would collapse. Invention, technology, industry—all of these exist in complete deference to nature, subject to its ordinary tests and sanctions, entropy, decay, extinction. Much that serves our purposes in the short term may work against us and the earth as we know it over time.

Human beings reshape the material world in ways that seem to meet their needs and desires. Needs, of course, are both biologically given and passed on by cultural tradition. The wiles of advertising exploit the fact that at the most ancient level, human needs and desires were shaped by the natural pressures and scarcities with which our ancestors lived. To the extent that the circumstances of human life have changed, through the exercise of human adaptive skills, the attempt to meet some needs may have become maladaptive.

This is the great and awful irony of "doing what comes naturally." The desire to have children is a product of past millennia when bands of human beings could barely keep up their numbers and as many as half of all offspring died too young to reproduce. High rates of infant survival are artifacts, not "natural" in the colloquial sense at all. Some religious groups reject contraception as "unnatural," yet the use of contraception to restore ancient balances is the use of artifact to repair the effects of artifact. The attempt to stave off death through biomedical technology is a similar result of desires that were once adaptive for the species. Because scarcity has been a fact of most of human existence, miserliness, overeating, and conspicuous consumption burden our lives today. Perhaps the delight in swift and powerful automobiles is a translation of the need to be able to run, whether in flight from predators or in pursuit of game. It's "natural" to want to own a gas-guzzling monster. It's "natural" to cling to life and the life of loved ones beyond any meaningful exchange or participation. The population explosion is "natural."

Most serious of all, the habit of seeing the human community as in some sense separate from (and in opposition to) nature is a natural habit, one that has appeared to be adaptive for our species through most of its history and may have ceased to be adaptive. Few cultures emphasize this separation as sharply as the Western tradition has, but even with stone-age technologies and the various mythologies of Earth kinship, the awareness is there.

The steady increase in the impact of the human species on all other species, on the atmosphere and the seas and the earth's surface, requires new patterns of adaptation and new kinds of perception, for the natural course of a species that destroys its environment is extinction. What we need to fashion today is a way of thinking that is both new and artificial—something deliberately dreamed up in the twentieth century and learned by all members of our species to protect the lives of future generations and preserve their options. We need to invent new forms and learn some new things: limits; moderation; fewer progeny; the acceptance of our own dying. We need to look further into the future, using more and better science and learning to think more clearly about our interdependence with other forms of life.

In doing so, we will be following our natures as the species that survives by learning.

MARY CATHERINE BATESON is Clarence Robinson Professor of Anthropology and English at George Mason University in Fairfax, Virginia. Bateson considers herself a self-styled "professional outsider" who has a unique gift for being able to discern recurrent abstract patterns in very different situations. She has written on a variety of linguistic and anthropological topics. She is the author of, among other books, *With a Daughter's Eye,* a memoir of her parents, Gregory Bateson and Margaret Mead; and *Composing a Life.*

Good and Bad Reasons
for Believing

Richard Dawkins

Dear Juliet,

Now that you are ten, I want to write to you about something that is important to me. Have you ever wondered how we know the things that we know? How do we know, for instance, that the stars, which look like tiny pinpricks in the sky, are really huge balls of fire like the sun and very far away? And how do we know that Earth is a smaller ball whirling round one of those stars, the sun?

The answer to these questions is "evidence." Sometimes evidence means actually seeing (or hearing, feeling, smelling . . .) that something is true. Astronauts have traveled far enough from Earth to see with their own eyes that it is round. Sometimes our eyes need help. The "evening star" looks like a bright twinkle in the sky, but with a telescope, you can see that it is a beautiful ball—the planet we call Venus. Something that you learn by direct seeing (or hearing or feeling . . .) is called an observation.

Often, evidence isn't just observation on its own, but ob-

servation always lies at the back of it. If there's been a murder, often nobody (except the murderer and the dead person!)
actually observed it. But detectives can gather together lots
of other observations which may all point toward a particular
suspect. If a person's fingerprints match those found on a
dagger, this is evidence that he touched it. It doesn't prove
that he did the murder, but it can help when it's joined up
with lots of other evidence. Sometimes a detective can think
about a whole lot of observations and suddenly realize that
they all fall into place and make sense if so-and-so did the
murder.

Scientists—the specialists in discovering what is true
about the world and the universe—often work like detectives. They make a guess (called a hypothesis) about what
might be true. They then say to themselves: *If* that were
really true, we ought to see so-and-so. This is called a prediction. For example, if the world is really round, we can
predict that a traveler, going on and on in the same direction,
should eventually find himself back where he started. When
a doctor says that you have measles, he doesn't take one look
at you and *see* measles. His first look gives him a *hypothesis*
that you *may* have measles. Then he says to himself: If she
really has measles, I ought to see . . . Then he runs through
his list of predictions and tests them with his eyes (have you
got spots?); hands (is your forehead hot?); and ears (does your
chest wheeze in a measly way?). Only then does he make his
decision and say, "I diagnose that the child has measles."
Sometimes doctors need to do other tests like blood tests or
X rays, which help their eyes, hands, and ears to make observations.

The way scientists use evidence to learn about the world
is much cleverer and more complicated than I can say in a
short letter. But now I want to move on from evidence,

which is a good reason for believing something, and warn you against three bad reasons for believing anything. They are called "tradition," "authority," and "revelation."

First, tradition. A few months ago, I went on television to have a discussion with about fifty children. These children were invited because they'd been brought up in lots of different religions. Some had been brought up as Christians, others as Jews, Muslims, Hindus, or Sikhs. The man with the microphone went from child to child, asking them what they believed. What they said shows up exactly what I mean by "tradition." Their beliefs turned out to have no connection with evidence. They just trotted out the beliefs of their parents and grandparents which, in turn, were not based upon evidence either. They said things like: "We Hindus believe so and so"; "We Muslims believe such and such"; "We Christians believe something else."

Of course, since they all believed different things, they couldn't all be right. The man with the microphone seemed to think this quite right and proper, and he didn't even try to get them to argue out their differences with each other. But that isn't the point I want to make at the moment. I simply want to ask where their beliefs came from. They came from tradition. Tradition means beliefs handed down from grandparent to parent to child, and so on. Or from books handed down through the centuries. Traditional beliefs often start from almost nothing; perhaps somebody just makes them up originally, like the stories about Thor and Zeus. But after they've been handed down over some centuries, the mere fact that they are so old makes them seem special. People believe things simply because people have believed the same thing over centuries. That's tradition.

The trouble with tradition is that, no matter how long ago a story was made up, it is still exactly as true or untrue as the original story was. If you make up a story that isn't true, handing it down over any number of centuries doesn't make it any truer!

Most people in England have been baptized into the Church of England, but this is only one of many branches of the Christian religion. There are other branches such as the Russian Orthodox, the Roman Catholic, and the Methodist churches. They all believe different things. The Jewish religion and the Muslim religion are a bit more different still; and there are different kinds of Jews and of Muslims. People who believe even slightly different things from each other go to war over their disagreements. So you might think that they must have some pretty good reasons—evidence—for believing what they believe. But actually, their different beliefs are entirely due to different traditions.

Let's talk about one particular tradition. Roman Catholics believe that Mary, the mother of Jesus, was so special that she didn't die but was lifted bodily into Heaven. Other Christian traditions disagree, saying that Mary did die like anybody else. These other religions don't talk about her much and, unlike Roman Catholics, they don't call her the "Queen of Heaven." The tradition that Mary's body was lifted into Heaven is not a very old one. The Bible says nothing about how or when she died; in fact, the poor woman is scarcely mentioned in the Bible at all. The belief that her body was lifted into Heaven wasn't invented until about six centuries after Jesus' time. At first, it was just made up, in the same way as any story like "Snow White" was made up. But, over the centuries, it grew into a tradition and people started to take it seriously simply *because* the story had been handed down over so many generations. The older the tradition became, the more people took it seriously. It finally was written down as an official Roman Catholic belief only very recently, in 1950, when I was the age you are now. But the story was no more true in 1950 than it was when it was first invented six hundred years after Mary's death.

I'll come back to tradition at the end of my letter, and look at it in another way. But first, I must deal with the two other bad reasons for believing in anything: authority and revelation.

Authority, as a reason for believing something, means believing it because you are told to believe it by somebody important. In the Roman Catholic Church, the pope is the most important person, and people believe he must be right just because he is the pope. In one branch of the Muslim religion, the important people are old men with beards called ayatollahs. Lots of Muslims in this country are prepared to commit murder, purely because the ayatollahs in a faraway country tell them to.

When I say that it was only in 1950 that Roman Catholics were finally told that they had to believe that Mary's body shot off to Heaven, what I mean is that in 1950, the pope told people that they had to believe it. That was it. The pope said it was true, so it had to be true! Now, probably some of the things that that pope said in his life were true and some were not true. There is no good reason why, just because he was the pope, you should believe everything he said any more than you believe everything that lots of other people say. The present pope has ordered his followers not to limit the number of babies they have. If people follow his authority as slavishly as he would wish, the results could be terrible famines, diseases, and wars, caused by overcrowding.

Of course, even in science, sometimes we haven't seen the evidence ourselves and we have to take somebody else's word for it. I haven't, with my own eyes, seen the evidence that light travels at a speed of 186,000 miles per second. Instead,

I believe books that tell me the speed of light. This looks like "authority." But actually, it is much better than authority, because the people who wrote the books have seen the evidence and anyone is free to look carefully at the evidence whenever they want. That is very comforting. But not even the priests claim that there is any evidence for their story about Mary's body zooming off to Heaven.

The third kind of bad reason for believing anything is called "revelation." If you had asked the pope in 1950 how he knew that Mary's body disappeared into Heaven, he would probably have said that it had been "revealed" to him. He shut himself in his room and prayed for guidance. He thought and thought, all by himself, and he became more and more sure inside himself. When religious people just have a feeling inside themselves that something must be true, even though there is no evidence that it is true, they call their feeling "revelation." It isn't only popes who claim to have revelations. Lots of religious people do. It is one of their main reasons for believing the things that they do believe. But is it a good reason?

Suppose I told you that your dog was dead. You'd be very upset, and you'd probably say, "Are you sure? How do you know? How did it happen?" Now, suppose I answered: "I don't actually know that Pepe is dead. I have no evidence. I just have this funny feeling deep inside me that he is dead." You'd be pretty cross with me for scaring you, because you'd know that an inside "feeling" on its own is not a good reason for believing that a whippet is dead. You need evidence. We all have inside feelings from time to time, and sometimes they turn out to be right and sometimes they don't. Anyway, different people have opposite feelings, so how are we to decide whose feeling is right? The only way to be sure that a dog is dead is to see him dead, or hear that his heart has stopped; or be told by somebody who has seen or heard some real evidence that he is dead.

People sometimes say that you must believe in feelings deep inside, otherwise, you'd never be confident of things like "My wife loves me." But this is a bad argument. There can be plenty of evidence that somebody loves you. All through the day when you are with somebody who loves you, you see and hear lots of little titbits of evidence, and they all add up. It isn't a purely inside feeling, like the feeling that priests call revelation. There are outside things to back up the inside feeling: looks in the eye, tender notes in the voice, little favors and kindnesses; this is all real evidence.

Sometimes people have a strong inside feeling that somebody loves them when it is not based upon any evidence, and then they are likely to be completely wrong. There are people with a strong inside feeling that a famous film star loves them, when really the film star hasn't even met them. People like that are ill in their minds. Inside feelings must be backed up by evidence, otherwise you just can't trust them.

Inside feelings are valuable in science, too, but only for giving you ideas that you later test by looking for evidence. A scientist can have a "hunch" about an idea that just "feels" right. In itself, this is not a good reason for believing something. But it can be a good reason for spending some time doing a particular experiment, or looking in a particular way for evidence. Scientists use inside feelings all the time to get ideas. But they are not worth anything until they are supported by evidence.

I promised that I'd come back to tradition, and look at it in another way. I want to try to explain why tradition is so important to us. All animals are built (by the process called evolution) to survive in the normal place in which their kind live. Lions are built to be good at surviving on the plains of Africa. Crayfish are built to be good at surviving in fresh water, while lobsters are built to be good at surviving in the salt sea. People are animals, too, and we are built to be good at surviving in a world full of . . . other people. Most of us don't hunt for our own food like lions or lobsters; we buy it from other people who have bought it from yet other people. We "swim" through a "sea of people." Just as a fish needs gills to survive in water, people need brains that make them able to deal with other people. Just as the sea is full of salt water, the sea of people is full of difficult things to learn. Like language.

You speak English, but your friend Ann-Kathrin speaks German. You each speak the language that fits you to "swim about" in your own separate "people sea." Language is passed down by tradition. There is no other way. In England, Pepe is a dog. In Germany he is *ein Hund.* Neither of these words is more correct, or more true than the other. Both are simply handed down. In order to be good at "swimming about in their people sea," children have to learn the language of their own country, and lots of other things about their own people; and this means that they have to absorb, like blotting paper, an enormous amount of traditional information. (Remember that traditional information just means things that are handed down from grandparents to parents to children.) The child's brain has to be a sucker for traditional information. And the child can't be expected to sort out good and useful traditional information, like the words of a language, from bad or silly traditional information, like believing in witches and devils and ever-living virgins.

It's a pity, but it can't help being the case, that because children have to be suckers for traditional information, they are likely to believe anything the grown-ups tell them, whether true or false, right or wrong. Lots of what the grown-ups tell them is true and based on evidence, or at least sensible. But if some of it is false, silly, or even wicked, there is nothing to stop the children believing that, too. Now, when the children grow up, what do they do? Well, of course, they tell it to the next generation of children. So, once something gets itself strongly believed—even if it is completely untrue and there never was any reason to believe it in the first place—it can go on forever.

Could this be what has happened with religions? Belief that there is a god or gods, belief in Heaven, belief that Mary never died, belief that Jesus never had a human father, belief that prayers are answered, belief that wine turns into blood—not one of these beliefs is backed up by any good evidence. Yet millions of people believe them. Perhaps this is because they were told to believe them when they were young enough to believe anything.

Millions of other people believe quite different things, because they were told different things when they were children. Muslim children are told different things from Christian children, and both grow up utterly convinced that they are right and the others are wrong. Even within Christians, Roman Catholics believe different things from Church of England people or Episcopalians, Shakers or Quakers, Mormons or Holy Rollers, and all are utterly convinced that they are right and the others are wrong. They believe different things for exactly the same kind of reason as you speak English and Ann-Kathrin speaks German. Both languages are, in their own country, the right language to speak. But it can't be true that different religions are right in their own countries, because different religions claim that opposite things are true. Mary can't be alive in Catholic Southern Ireland but dead in Protestant Northern Ireland.

What can we do about all this? It is not easy for you to do anything, because you are only ten. But you could try this. Next time somebody tells you something that sounds important, think to yourself: "Is this the kind of thing that people probably know because of evidence? Or is it the kind of thing that people only believe because of tradition, authority, or revelation?" And, next time somebody tells you that something is true, why not say to them: "What kind of evidence is there for that?" And if they can't give you a good answer, I hope you'll think very carefully before you believe a word they say.

Your loving

Daddy

RICHARD DAWKINS is an evolutionary biologist; reader in the Department of Zoology at Oxford University; fellow of New College. He began his research career in the 1960s as a research student with Nobel Prize–winning ethologist Nico Tinbergen, and ever since then, his work has largely been concerned with the evolution of behavior. Since 1976, when his first book, *The Selfish Gene,* encapsulated both the substance and the spirit of what is now called the sociobiological revolution, he has become widely known, both for the originality of his ideas and for the clarity and elegance with which he expounds them. A subsequent book, *The Extended Phenotype,* and a number of television programs, have extended the notion of the gene as the unit of selection, and have applied it to biological examples as various as the relationship between hosts and parasites and the evolution of cooperation. His following book, *The Blind Watchmaker,* is widely read, widely quoted, and one of the truly influential intellectual works of our time. He is also author of the recently published *River Out of Eden.*

Part Two

Origins

What Happened Before the Big Bang?

Paul Davies

Well, what did happen before the big bang?

Few schoolchildren have failed to frustrate their parents with questions of this sort. It often starts with puzzlement over whether space "goes on forever," or where humans came from, or how the planet Earth formed. In the end, the line of questioning always seems to get back to the ultimate origin of things: the big bang. "But what caused *that*?"

Children grow up with an intuitive sense of cause and effect. Events in the physical world aren't supposed to "just happen." Something makes them happen. Even when the rabbit appears convincingly from the hat, trickery is suspected. So could the entire universe simply pop into existence, magically, for no actual reason at all?

This simple, schoolchild query has exercised the intellects of generations of philosophers, scientists, and theologians. Many have avoided it as an impenetrable mystery. Others have tried to define it away. Most have got themselves into an awful tangle just thinking about it.

The problem, at rock bottom, is this: If nothing happens without a cause, then *something* must have caused the universe to appear. But then we are faced with the inevitable question of what caused that *something*. And so on in an infinite regress. Some people simply proclaim that God created the universe, but children always want to know who created God, and *that* line of questioning gets uncomfortably difficult.

One evasive tactic is to claim that the universe didn't *have* a beginning, that it has existed for all eternity. Unfortunately, there are many scientific reasons why this obvious idea is unsound. For starters, given an infinite amount of time, anything that can happen will already have happened, for if a physical process is likely to occur with a certain nonzero probability—however small—then given an infinite amount of time the process *must* occur, with probability one. By now, the universe should have reached some sort of final state in which all possible physical processes have run their course. Furthermore, you don't *explain* the existence of the universe by asserting that it has always existed. That is rather like saying that nobody wrote the Bible: it was just copied from earlier versions. Quite apart from all this, there is very good evidence that the universe *did* come into existence in a big bang, about fifteen billion years ago. The effects of that primeval explosion are clearly detectable today—in the fact that the universe is still expanding, and is filled with an afterglow of radiant heat.

So we are faced with the problem of what happened beforehand to trigger the big bang. Journalists love to taunt scientists with this question when they complain about the money being spent on science. Actually, the answer (in my opinion) was spotted a long time ago, by one Augustine of Hippo, a Christian saint who lived in the fifth century. In those days before science, cosmology was a branch of theology, and the taunt came not from journalists, but from pagans: "What was God doing before he made the universe?" they asked. "Busy creating Hell for the likes of you!" was the standard reply.

But Augustine was more subtle. The world, he claimed, was made "not *in* time, but simultaneously *with* time."

In other words, the origin of the universe—what we now call the big bang—was not simply the sudden appearance of matter in an eternally preexisting void, but the coming into being of time itself. Time *began* with the cosmic origin. There was no "before," no endless ocean of time for a god, or a physical process, to wear itself out in infinite preparation.

Remarkably, modern science has arrived at more or less the same conclusion as Augustine, based on what we now know about the nature of space, time, and gravitation. It was Albert Einstein who taught us that time and space are not merely an immutable arena in which the great cosmic drama is acted out, but are part of the cast—part of the physical universe. As physical entities, time and space can change—suffer distortions—as a result of gravitational processes. Gravitational theory predicts that under the extreme conditions that prevailed in the early universe, space and time may have been so distorted that there existed a boundary, or "singularity," at which the distortion of space-time was infinite, and therefore through which space and time cannot have continued. Thus, physics predicts that time was indeed bounded in the past as Augustine claimed. It did not stretch back for all eternity.

If the big bang was the beginning of time itself, then any

discussion about what happened before the big bang, or what caused it—in the usual sense of physical causation—is simply meaningless. Unfortunately, many children, and adults, too, regard this answer as disingenuous. There must be more to it than that, they object.

Indeed there is. After all, *why* should time suddenly "switch on"? What explanation can be given for such a singular event? Until recently, it seemed that any explanation of the initial "singularity" that marked the origin of time would have to lie beyond the scope of science. However, it all depends on what is meant by "explanation." As I remarked, all children have a good idea of the notion of cause and effect, and usually an explanation of an event entails finding something that caused it. It turns out, however, that there are physical events which do *not* have well-defined causes in the manner of the everyday world. These events belong to a weird branch of scientific inquiry called quantum physics.

Mostly, quantum events occur at the atomic level; we don't experience them in daily life. On the scale of atoms and molecules, the usual commonsense rules of cause and effect are suspended. The rule of law is replaced by a sort of anarchy or chaos, and things happen *spontaneously*—for no particular reason. Particles of matter may simply pop into existence without warning, and then equally abruptly disappear again. Or a particle in one place may suddenly materialize in another place, or reverse its direction of motion. Again, these are real effects occurring on an atomic scale, and they can be demonstrated experimentally.

A typical quantum process is the decay of a radioactive nucleus. If you ask why a given nucleus decayed at one particular moment rather than at some other, there is no answer. The event "just happened" at that moment, that's all. You cannot predict these occurrences. All you can do is give the probability—there is a fifty-fifty chance that a given nucleus will decay in, say, one hour. This uncertainty is not simply a result of our ignorance of all the little forces and influences that try to make the nucleus decay; it is inherent in nature itself, a basic part of quantum reality.

The lesson of quantum physics is this: Something that "just happens" need not actually violate the laws of physics. The abrupt and uncaused appearance of something can occur within the scope of scientific law, once quantum laws have been taken into account. Nature apparently has the capacity for genuine spontaneity.

It is, of course, a big step from the spontaneous and un-caused appearance of a subatomic particle—something that is routinely observed in particle accelerators—to the spontaneous and uncaused appearance of the universe. But the loophole is there. If, as astronomers believe, the primeval universe was compressed to a very small size, then quantum effects must have once been important on a cosmic scale. Even if we don't have a precise idea of exactly what took place at the beginning, we can at least see that the origin of the universe from nothing need not be unlawful or unnatural or unscientific. In short, it need not have been a supernatural event.

Inevitably, scientists will not be content to leave it at that. We would like to flesh out the details of this profound concept. There is even a subject devoted to it, called quantum cosmology. Two famous quantum cosmologists, James Hartle and Stephen Hawking, came up with a clever idea that goes back to Einstein. Einstein not only found that space and time are part of the physical universe; he also found that they are linked in a very intimate way. In fact, space on its own and time on its own are no longer properly valid concepts. Instead, we must deal with a unified "space-time" continuum. Space has three dimensions, and time has one, so space-time is a four-dimensional continuum.

In spite of the space-time linkage, however, space is space and time is time under almost all circumstances. Whatever space-time distortions gravitation may produce, they never turn space into time or time into space. An exception arises, though, when quantum effects are taken into account. That all-important intrinsic uncertainty that afflicts quantum systems can be applied to space-time, too. In this case, the uncertainty can, under special circumstances, affect the *identities* of space and time. For a very, very brief duration, it is possible for time and space to merge in identity, for time to become, so to speak, spacelike—just another dimension of space.

The spatialization of time is not something abrupt; it is a continuous process. Viewed in reverse as the temporalization of (one dimension of) space, it implies that time can *emerge out of space* in a continuous process. (By continuous, I mean that the timelike quality of a dimension, as opposed to its spacelike quality, is not an all-or-nothing affair; there are shades in between. This vague statement can be made quite precise mathematically.)

The essence of the Hartle-Hawking idea is that the big bang was not the abrupt switching on of time at some singular first moment, but the emergence of time from space in an ultrarapid but nevertheless continuous manner. On a human time scale, the big bang was very much a sudden, explosive origin of space, time, and matter. But look very, very closely at that first tiny fraction of a second and you find that there was no precise and sudden beginning at all. So here we have a theory of the origin of the universe that seems to say two contradictory things: First, time did not always exist; and second, there was no first moment of time. Such are the oddities of quantum physics.

Even with these further details thrown in, many people feel cheated. They want to ask *why* these weird things happened, *why* there is a universe, and why *this* universe. Perhaps science cannot answer such questions. Science is good at telling us how, but not so good on the why. Maybe there isn't a why. To wonder why is very human, but perhaps there is no answer in human terms to such deep questions of existence. Or perhaps there is, but we are looking at the problem in the wrong way.

Well, I didn't promise to provide the answers to life, the universe, and everything, but I have at least given a plausible answer to the question I started out with: What happened before the big bang?

The answer is: Nothing.

PAUL DAVIES is a theoretical physicist and professor of natural philosophy at the University of Adelaide. He has published over one hundred research papers in the fields of cosmology, gravitation, and quantum field theory, with particular emphasis on black holes and the origin of the universe. He is also interested in the nature of time, high-energy particle physics, the foundations of quantum mechanics, and the theory of complex systems. He runs a research group in quantum gravity which is currently investigating superstrings, cosmic strings, higher-dimensional black holes, and quantum cosmology.

Davies is well known as an author, broadcaster, and public lecturer. He has written over twenty books, ranging from specialist textbooks to popular books for the general public. Among his better-known works are *God and the New Physics; Superforce; The Cosmic Blueprint;* and *The Mind of God.* His most recent books are *The Last Three Minutes* and *It's About Time.*

He was described by the *Washington Times* as "the best science writer on either side of the Atlantic." He likes to focus on the deep questions of existence, such as how the universe came into existence and how it will end, the nature of human consciousness, the possibility of time travel, the relationship between physics and biology, the status of the laws of physics, and the interface of science and religion.

The Joy of Water

P. W. Atkins

There is much joy to be had from water. It is not merely its awesome abundance and the variety of its forms that are so joy inspiring, or even its essential role in the carving of our planet and the evolution of life. For me, the joy is that such rich properties can emerge from such a simple structure. Moreover, it is not merely the richness of the simplicity of water that is such an inspiration, for I also find deep satisfaction in the subtlety of its properties. That these unusual properties are crucial to the emergence and persistence of life is a dimension added to the pleasure of beholding water.

Water—be it the rolling Pacific Ocean or a droplet of morning mist, a craggy glacier of a snowflake, a gas pulsing through the blades of a steam turbine or hanging in the air as a major contribution to the global turmoil we call the weather—is composed of water molecules. Every water molecule in the world, and wherever else it occurs in the universe, is identical. Each one consists of a central oxygen atom to which are attached two hydrogen atoms. That is all. Oceans, life, and romance all stem from this simple picture.

To see the potential for the formation of the Pacific Ocean in this minuscule entity, we need to know some details about hydrogen, the lightest element of all, and its companion element, oxygen. Atoms of hydrogen are very small; they consist of a single positively charged proton for a nucleus, and that proton is surrounded by a single electron. One advantage of being small is that the nucleus of such an atom can penetrate close to the electrons of other atoms. Hydrogen nuclei can wriggle into regions that bigger atoms cannot reach. Moreover, because there is only one electron in a hydrogen atom, the bright positive charge of the nucleus can blaze through the cloudlike electron's misty negative charge, and there is, consequently, a strong attraction for other electrons that happen to be nearby.

As for the oxygen atom, it is much bigger than a hydrogen atom. Nevertheless, as atoms go, it is still quite small compared to those of other elements, such as sulfur, chlorine, and even carbon and nitrogen. Its smallness stems from the strong positive charge of the oxygen nucleus, which draws its electron close to itself. Additionally, because it is so small and yet has a strongly charged nucleus, an oxygen atom can draw toward itself the electrons of other atoms. In particular, it can draw in the electrons of any atoms to which it is bonded.

In water, an oxygen atom is bonded to two small hydrogen atoms. The central oxygen atom sucks in the electrons of the oxygen-hydrogen links, and so partially denudes the hydrogen still further of their electrons. The oxygen atom thereby becomes oxygen rich and the hydrogen atoms become electron poor. Consequently, the oxygen atom has a vestigial negative charge (arising from its being bloated with electrons) and the hydrogen atoms acquire a vestigial positive charge because the positive charge of the nucleus is no longer canceled by the surrounding electrons (for they have been partially sucked away). The resulting distribution of charge—oxygen negative and hydrogen positive—coupled with the small size of the hydrogen atoms, is at the root of water's extraordinary properties.

Another feature that conspires with the distribution of electrons and results in oceans is the shape of the water molecule. It is an angular, open-V-shaped molecule, with the oxygen atom at the apex of the V. The important feature of this shape, which can be rationalized by examining how electrons are arranged around the central oxygen atom, is that one side of the oxygen atom is exposed, and that exposed side is rich in electrons.

Now we shall see how these features bubble out into the real world of phenomena and tangible properties. Most important of all is the ability of one water molecule to stick to another water molecule. The electron-rich region of the oxygen atom is the site of negative electric charge; the partially denuded hydrogen atom of a neighboring molecule is the site of positive charge, and the opposite charges attract one another. The special link between the two water molecules mediated by a hydrogen atom in this way is called a *hydrogen bond.* It is one of the most important intermolecular links in the world, for its effects range from the operation of the genetic code (the two strands of the DNA double helix are linked together by hydrogen bonds), through the toughness of wood (for the ribbons of cellulose are clamped rigidly together face-to-face by the sturdy and numerous hydrogen bonds between them), and—the point of our concern—with the properties of water. For a water molecule is so light that if it were not for the hydrogen bonds that can form between its molecules, then water would be a gas, and instead of puddles, lakes, and oceans of precious liquid, there would be a humid sky full of gaseous water and barren ground beneath.

Just as hydrogen bonds between water molecules trap them into forming a liquid even at warm everyday temperatures, so they also help to form the rigid solid ice at only slightly lower temperatures. However, when ice forms from liquid water, something rather odd happens, an oddness that is also life preserving. When the temperature is lowered, the water molecules of a liquid are shaken and jostled less vigorously, and hydrogen bonds can form more extensively and survive for longer. As a result, the molecules cease flowing readily as a liquid, and a stable solid forms instead. Now the shape of the molecule comes into play. An oxygen atom in the V-shaped water molecule has room to accommodate two hydrogen bonds, one to each of two neighboring molecules. Each oxygen atom now participates in four bonds—two ordinary oxygen-hydrogen bonds, and two hydrogen bonds to neighbors—and these four bonds point toward the corners of a tetrahedron. This arrangement, which is continued neighbor after neighbor through the solid, results in a very open structure for ice, and the water molecules are held apart as well as held together, like an open scaffold of atoms and bonds. When ice melts, this open structure collapses and forms a denser liquid. When water freezes, the collapsed structure of the liquid unfurls and expands into an open structure.

In other words, almost uniquely among substances, the solid form (ice) is less dense than the liquid form. One consequence of this peculiarity is that ice forms and floats on the surface of lakes. This feature is life preserving, because the film of ice helps to protect the water below from the freezing effect of the air above, and marine life can survive and flourish even though the temperature is low enough to freeze the surface layers of water.

The hydrogen bonds in water and the tightness with which they bind molecule to molecule are also responsible for other features of water. The color of water in bulk, which gives our planet its singular hue, can be traced to them; so can the film that forms on the surface of the liquid and which curves it into droplets. Water's considerable heat capacity (its ability to store energy supplied as heat) is another consequence of these bonds, and this characteristic is put to use in domestic central heating systems, where a little water can be used to pump a great deal of energy around a house.

Another extraordinary feature of water is its ability to dissolve so much. This characteristic also stems from the peculiar arrangement of electric charges and atoms in a water molecule. Many compounds consist of ions, or electrically charged atoms. Common salt, sodium chloride, for instance, consists of positively charged sodium ions and negatively charged chloride ions. In the solid, each positive ion is surrounded by negative ions, and each negative ion is surrounded by positive ions. Water, though, with its system of positive and negative charges, can emulate both types of these surrounding ions. Thus, when exposed to water, the sodium ions of a crystal can become surrounded by water molecules that present their negatively charged oxygen atoms toward them, and thereby emulate chloride ions. Similarly, chloride ions can become surrounded by positively charged hydrogen atoms of the water molecules, which emulate the effect of sodium ions in the original crystal. Each type of ion is seduced; the sodium ions float off surrounded by water molecules emulating chloride ions, and the chloride ions float off surrounded by water molecules using their hydrogen atoms to emulate sodium ions. Water has a peculiarly strong ability to act in this way, which is why it is such a good solvent (or, in some circumstances, when it corrodes, a highly dangerous chemical). It is this ability of water that carves landscapes from stone. It transports nutrients through the soil and brings them into plants. Water pervades our bodies, and through its ability to support the free motion of ions and other molecules that it dissolves, provides an environment for life.

Water is truly a remarkable substance; so slight in structure, yet so huge in physical and chemical stature. That so meager an entity can behave so grandly is a microcosm of modern science, which seeks giants of simplicity and thereby adds joy to our appreciation of this wonderful world.

PETER ATKINS has been a Fellow of Lincoln College, Oxford University; university lecturer in physical chemistry since 1965; and visiting professor in a number of institutions, including universities in France, Japan, China, New Zealand, and Israel. He was awarded the Meldola Medal of the Royal Society of Chemistry in 1969, and an honorary degree from the University of Utrecht for his contributions to chemistry.

His books include *Physical Chemistry; Inorganic Chemistry;* and *General Chemistry.* In addition, he has written a number of books on science for the general public, including *The Second Law; Molecules;* and *Atoms, Electrons, and Change.* His interests extend to cosmology and the deep contribution of science to culture. His overriding focus is on the communication of science, and he seeks to share the thrill and pleasure that scientific insights provide.

Where Do We Come From?

Robert Shapiro

~~~~~~~~~~ **M**y son had many things to play with during his childhood, but one was special. Her name was Frizzle. She was a ball of inquisitive fur several inches long that the pet store classified as a gerbil. Frizzle spent much of her time in the brief few years of her life in exploring the multichambered residence that we had prepared for her, and then in trying to escape from it. When she died, we missed her.

For various reasons, I had no animal pet when I was young, but for a time, I attempted to nourish a small cactus plant. Its activities were less interesting than those of a gerbil; it grew slightly, but made no effort to escape. Yet I was saddened when it turned gray and drooped, and I realized that it had lost the struggle for survival.

We all learn at an early age how profoundly living things change when they die. We also recognize that the living things we know are a small part of a larger surrounding universe of things like water, rocks, and the moon that are not, and have never been, alive. This wisdom belongs to our own time, however. For centuries, many observers, including skilled scientists, did not recognize that dead things do not become alive. They felt, for example, that river mud could give rise to serpents, and raw meat could give birth to worms in a process called spontaneous generation. Only through many carefully controlled experiments, culminating in a brilliant series carried out by Louis Pasteur in the nineteenth century, was this theory disproven. We now recognize that life comes only from previously existing life, like a flame that can be divided, and spread, but once extinguished, can never be rekindled.

How then did life first come into existence, on this planet or anywhere else that life may exist in the universe? Many religions and some philosophies avoid the problem by presuming that life, in the form of a deity or other immortal being, has existed eternally. An alternative view can be found in science, where we look for natural answers in preference to supernatural ones and turn to another possibility: that life arose from nonlife at least once, sometime after the creation of the universe.

To learn about life's origins, we obviously cannot rely upon human records or memories, and must turn instead to the evidence stored in the earth itself. Those data are left as fossils in sediments whose age can be deduced from the amount of radioactivity remaining in surrounding rocks. For example, an unstable isotope of potassium is sealed into a volcanic rock when it solidifies from lava. Half of this isotope decays every 1.3 billion years, with part of it converted to the stable gas argon, which remains trapped in the rocks. By measuring the amounts of the remaining potassium isotope and the trapped argon, and performing a simple calculation, we can determine the age of the rock.

The fossil record tells a marvelous tale of the rise of life. It starts three and a half billion years ago with single-celled forms that resemble bacteria and algae, and leads up to the overwhelming variety of living creatures, including ourselves, that are present today. Earth itself is only one billion years older than the earliest record of life discovered thus far. If we were to compare Earth's age to that of a sixty-year-old man, then the fraction of Earth's existence needed for life to appear could be compared to the time needed for the man to reach puberty. The geological time occupied by all recorded human history would be equivalent to the last half-hour or so of the man's existence. Unfortunately, the record of geology peters out beyond three and a half billion years ago. No rocks remain that tell us anything about the way in which the first bacteria-like creatures came into existence. This process remains a profound puzzle.

A human being is so much more than a bacterium, though. If we can understand the evolutionary process that produced ourselves from one-celled creatures, and the developmental changes that convert a one-celled fertilized egg into a multicelled adult, then why is it difficult to understand how a bacteria could form from nonliving matter?

To appreciate this problem, we must explore the structure of life as it exists today. Obviously, this topic can fill many volumes, but here I wish to draw out a single thread that serves to distinguish life from nonlife: Living things are highly organized. I have chosen a word that is familiar; some scientists prefer terms like *negative entropy,* but basically, the idea is the same. By using *organized,* I mean to describe the quality that distinguishes the complete works of William Shakespeare from a series of letters struck at random on a typewriter, or a symphony from the sound one gets by dropping dishes on the floor. Clearly, things produced by the activities of life, for example, Shakespeare's works, can also be organized, but things unconnected with life, like a rock on the surface of the moon, are much less so. A bacterium, when compared with nonliving matter, is still very organized. The comparison between a dust particle and a bacterium would then be similar to the match of a random string of letters to a Shakespeare play.

Some scientists who have thought about life's origins have believed that this gap in organization between living and nonliving matter could be closed by random chance alone, if enough tries were made. They have drawn great encouragement from a famous experiment devised by Stanley Miller and Harold Urey. Miller and Urey showed in 1953 that certain amino acids could be formed very easily when electrical energy is passed through a simple mixture of gases. These amino acids are building blocks of proteins, one of life's key components. If key chemicals can arise so readily, can the rest of life be far behind?

Unfortunately, life is vastly more organized than the "prebiotic" chemical mixtures formed in Miller-Urey–type experiments. Imagine that by random strokes on the typewriter you type the words *to be*. You might then be reminded of the famous line: "To be, or not to be: that is the question." A further jump of imagination might lead you to the idea that the remainder of *Hamlet* could then emerge from the random strokes. But any sober calculation of the odds reveals that the chances of producing a play or even a sonnet in this way are hopeless, even if every atom of material on Earth were a typewriter that had been turning out text without interruption for the past four and a half billion years.

Other thinkers, including a number of religious people, have argued that the formation of life from nonliving matter by natural means is impossible. They cite the Second Law of

Thermodynamics, and claim that the formation of organized matter from disorganized matter is forbidden by it. But the Second Law applies only to sealed-in (closed) systems. It does not forbid nonliving chemicals on Earth to absorb energy from an outside source such as the sun and becoming more organized. The gain in organization here would be balanced by a greater loss of organization in the sun, and the Second Law would thus be satisfied.

When chemical systems absorb energy, however, they usually use it to heat themselves up, or to form new bonds in ways that do not lead to any gain in organization. We don't know the key recipe—the set of special ingredients and forms of energy that could lead chemical systems up the ladder of organization on the first steps to life. These circumstances may be quite rare and difficult or, once you have grasped the trick, as simple as brewing beer or fermenting wine.

How can we find out more? One way would be to run more prebiotic experiments. Many have been run, of course, but they usually have searched for the chemicals present in life today, rather than seeking to identify the process of self-organization. It isn't likely that the highly evolved biochemicals of today—proteins, nucleic acids, and other complexities—were present during the first faltering steps to life. What we need is more understanding of how simple chemical substances such as minerals, soaps, and components of the air will behave when exposed to an abundant and continuing supply of energy, such as ultraviolet light. Would the material simply turn into a tar or dissipate the energy as heat? Most of the time this would happen, and little of importance would have been learned. But perhaps, if the right mixture were chosen, complex chemical cycles would establish themselves and continue to evolve. If so, we would have gained an important clue about the beginning of life. Some experiments of this type could be carried out in undergraduate and even high-school laboratories, as they would not require complex and costly equipment. This area remains one where amateurs could make a significant contribution to fundamental science.

Another scientific approach to seeking our origins is very expensive, but it would provide great excitement and inspiration along the way. In the past generation, we have developed the ability to explore our solar system, but not good enough reasons to ensure that this is done. By that I mean objectives that would capture the attention of the public and motivate them to support the costs, rather than ones of interest only to scientists deeply immersed in their specialties. The solar system offers a dazzling array of worlds, each containing different chemical systems that have been exposed to energy for billions of years. Some of them may have developed in the direction of organization. By discovering a system that has started on such a path, even a different path from the one followed on our own planet, we may get vital clues about the principles involved in self-organization and the nature of our own first steps. A treasure hunt of this type among the worlds surrounding our sun might or might not discover early forms of life, but would certainly put some life back into our space program.

"Where do we come from?" In my title, I put the origin of life in terms of a question of location, as a child might do. Some scientists have argued that life started elsewhere, then migrated to planet Earth. Even if this were so, that fact would still not solve the central question, which is one of mechanism: "How did we come to be?" Location may be critical, though, in another way: To learn how we started, even if it happened here, we may have to venture out into the greater universe that awaits us.

ROBERT SHAPIRO is professor of chemistry at New York University. He is author or coauthor of over ninety articles, primarily in the area of DNA chemistry. In particular, he and his coworkers have studied the ways in which environmental chemicals can damage our hereditary material, causing changes that can lead to mutations and cancer. His research has been supported by numerous grants from the National Institutes of Health; the Department of Energy; the National Science Foundation; and other organizations.

In addition to his research, Professor Shapiro has written three books for the general public. The topics have included the extent of life in the universe (*Life Beyond Earth,* with Gerald Feinberg); the origin of life on earth (*Origins: A Skeptic's Guide to the Creation of Life on Earth*); and the current effort to read the human genetic message (*The Human Blueprint*).

Who Do We Blame for What We Are?

Jack Cohen

⁓⁓⁓⁓⁓ Look at a fly. No, don't just imagine one, actually find a real fly and watch it. Watch its little legs run in perfect rhythm, its little head turn to watch events. Watch it take off: One moment it's standing there, the next it's in the air, zooming about and not bumping into things. If you can, watch it *land,* that's the most impressive trick: It zooms toward a wall, tilts, brakes, and . . . there it is on the surface, totally unruffled, cleaning its proboscis with its front legs. Where did all that beautifully precise machinery come from?

How did this fly, this one individual, come to be? It came from a fly's egg, you say. Not quite. It came from a maggot, and the maggot came out of an egg. We're so used to complicated things coming out of eggs that this sounds like an explanation. Fluffy little chicks come out of featureless oval (of course) chicken eggs. All you have to add is some warmth. *You* came from an egg, too.

But eggs are rather uncomplicated biological structures.

Compared with what comes out of them, they really are quite simple. In a fertile hen's egg, the dozens of cells on the yolk, from which the chick will arise, are genuinely trivial compared with a tiny bit of the chick's brain, its kidney, even its skin. One growing feather is, in all the ways that we can measure, more complicated than the cells from which the whole chick originated. How can this be? How can complication arise from simplicity? Is there an "organizing principle," a "Spirit of Life"?

The usual answer, today, is that there is. It's the DNA blueprint, great long molecules with enormous amounts of information written along them in a language of four nucleotide letters, in the nucleus of each of the cells. This is supposed to tell the developing organism—fly, chick, or you—how to build itself. According to this simplistic view of DNA and development, the organism is the DNA information made flesh. Fly DNA makes flies; chicken DNA makes chickens; people DNA makes people. But this isn't really how DNA functions in development, although you might not understand that from popular accounts or even from some biology textbooks.

There are several ways of looking at DNA. A solution of human DNA, chopped into moderately short lengths, is a sticky liquid in a test tube. In each tiny human nucleus, there are nearly two yards of DNA thread. If a human-cell nucleus were magnified a thousand times—it would then be the size of an aspirin tablet—it would contain a mile of DNA wire packed into it! Obviously, there's a lot of room for a lot of instructions, a lot of blueprints, on such a long strand of DNA. But DNA doesn't work to get this information turned into flies and chickens and people. It just sits there. Like the recipes for a wonderful meal in the pages of the cookbook on the shelf, it just sits there.

How then does the DNA "make" the fly? The straight answer is that it doesn't. But it's a difficult bit of thinking to realize why not, and that is why most people think it does—and that all you need to get a dinosaur is some dinosaur DNA. Dinosaurs and flies aren't made of DNA, just as veal *cordon-bleu* isn't made from the paper and ink of the cookbook.

Bacterial DNA is easy to think about. Bacteria are like simple workshops, full of chemical tools. Some of these tools read along the DNA thread; others make more tools according to what is read on the thread (including the tools that read the thread). Others are structural, or are chemical pumps, or are concerned with food or energy. The little "workshop" is actively producing more of its own parts, tools, bricks for its walls; it is growing. Some of the tools duplicate the DNA—like copying a cassette tape, in many ways—and surpluses of other tools accumulate. Then particular tools arrange for a division, and the process continues in the two daughter bacteria.

An egg is rather more complicated than a bacterium (but much less than what comes from it). It's made of the same kind of stuff, many of the tools are the same—but the life processes are completely different. It doesn't just grow and divide, it develops; it becomes something else. As a result of what the early egg's tools do, it turns into a different, usually bigger, more complicated structure called an embryo. The embryo uses yolk as a source of energy and building material, and makes a different kind of thing again. Perhaps it's a larva like the maggot of the fly, or a human fetus which then develops into a baby. The bacterial workshop only makes more of the same—but the egg makes new kinds of tools and equipment at each new stage. When the fly egg produces a maggot that feeds itself, this is like the bacterial "workshop" turning itself into a truck to go down to the hardware store to get what it needs to work—and to turn into the next stage. Similarly, the human embryo makes a placenta to get food and energy from mother's blood so it can build itself into the next stage. Let's take this analogy further for the maggot, to show you how miraculous development really is. The maggot truck builds itself bigger and bigger, then it drives off to a quiet country lane and builds itself a little garage (the pupa). Inside this it reassembles itself. Out comes—an airplane, the adult fly. A tiny, self-propelled, robot-controlled, self-fueling airplane—and half of the flies have little egg workshops inside, ready to start the process again.

DNA is involved all the time, to specify the tools to be used. Some DNA sequences—genes—specify biochemical tools which are like clocks, clamps, lathes, workbenches, worksheets, and schedules that regulate how the work is done. But the DNA does not have a fly description, or a chick pattern or a "you" pattern. It doesn't even have a wing pattern or a nose pattern. It is much more useful to think of each character (like a nose) being contributed to by *all* the DNA genes, and each DNA gene contributing to all characters, than it is to think of each character having its own little set of genes that "make" it. However, we can (in principle, and nearly in practice) list all of a fruit fly's genes and mark those which, when they mutate, cause changes (usually problems) with the wings. The temptation that many geneticists, and nearly all media reporters, fall into is to think that these genes are the fruit fly's "wing kit." It isn't a "wing kit," because nearly all of the genes affect other things, too. One of the *vestigial-wing* mutations, for example, damages the working of a molecular pump present in all cells. One of its many effects is that the wing can't be inflated properly when the fly comes out of the pupa. There are some genes, called "homeotic" genes, that have more radical, yet more specific, results. Changes to homeotic DNA sequences can alter the specification of an organ, and produce a different one: the *antennapedia* mutation replaces an antenna by a leg; *cockeyed* replaces eyes by genital structures. These are gene sequences which, in our workshop analogy, specify the geography of the workshop—where things should be built. Homeotic mutations change colors on the embryonic map so that the cells in the antenna rudiment "think" that they are in leg-rudiment positions—so they make good legs, but in the wrong place.

There is much more, of course, that the DNA doesn't have to specify—or that it can't change. It doesn't have to make water wet, or make fats unwettable, or make sodium chloride crystals cubical (but it can change the freezing point of watery solutions by making antifreeze proteins). There are many physical and chemical "givens." There are many biological regularities that are nearly as "given" as these physical and chemical mechanisms. A very important, and very ancient, DNA tool kit copies DNA with almost perfect fidelity; and some long-established energy-exchange mechanisms are common to many kinds of living things, too. As much as 60 percent of the informational DNA consists of these "conserved sequences," the same in the fly, the chicken, and human beings. Indeed, many of these basic "housekeeping" genes are the same in oak trees and bacteria, too. Therefore, most of what goes on to make a fly is much the same as what goes on to make you.

Why, then, are organisms so different? Let us be more imaginative, briefly. Differences between organisms, *however large,* need not be due to a major difference in the DNA. In principle, just one tiny difference could switch development into a new path, and even the fly and the chicken *could* have identical DNA except for that one early switch at which the paths diverge. The developmental difference, the imaginary fly/chicken switch, needn't even be at DNA level. If chickens like it hot, and incubate their eggs, and flies like it cold— then the chicken program could reproduce chickens with high-temperature development, and the fly program could reproduce flies at low temperatures. They could, in principle, use precisely the same DNA. That kind of "thought experiment" demonstrates that you simply can't tell which organism a particular DNA kit will "make," or which DNA will "make" a particular organism. As the mathematicians say, there is no "mapping" from DNA sequence to the structure of the organism whose development it contributes to.

The DNA in the egg can't start development, either, by itself; the reading-DNA-and-acting tool kits must be in working order, and actually working. They are provided by the rest of the egg, around the original nucleus. Development of nearly all animals is begun in the ovary of the mother, as the egg cells are constructed. Even the development of the embryo doesn't need the offspring's DNA messages until the egg structure has made the basic architecture of the future animal. Only then do homeobox genes "know where they are" and what they should do. In a way, fertilization happens quite late in development. The egg has been quite ready to get on with making an animal, and the sperm acts as a trigger only (apart from its donation of DNA, which differs little from that of the egg). The egg is like a loaded gun. A better mechanical analogy is to see the nonnuclear parts of the egg as a tape player, and the nuclear DNA as the tape. The first steps in development involve putting tape into the right slot for it to be played; setting the volume, speed, and so on; choosing the tracks to play and in what order; then pressing PLAY.

To continue our fanciful "two different animals with the same DNA" imaginations in the last paragraph, we could have the animals with a small difference in the egg tape-player mechanism: Fly eggs could read the DNA genes in a certain order (say, a, b, c, d) and make a fly whose ovary made eggs that did this again; while chicken eggs might read the tape z, y, x, w—and result in a chicken. If we swapped chicken for fly DNA in this hypothetical case, it would make no difference, because we've imagined that they're the same DNA. Even the different ovaries (like all the other organs) would be consistently different—and both organisms would breed true.

We can use a converse argument to show how absurd is the idea that DNA "contains the instructions for making the animal." Let's put fly DNA into a chicken egg (real fly DNA, not imaginary fly=chicken DNA). Even if the tape could be read at all, systematically enough that an embryo could result, what would we get by continuing the development of a fly using the basic structure of a bird? If a miracle occurred, even, and we got a functioning maggot—how would the fly get out of the shell? The reverse is worse: Even if the little chick embryo could start making itself with the fly beginnings, it would soon run out of yolk, and the tiny chick embryo wouldn't be much good at finding food for further development. So dinosaur DNA, from dinosaur blood preserved in a tick in amber, *can't* make a dinosaur. To play a dinosaur DNA tape, you need a dinosaur egg of the same species: the right tape player. DNA is not enough, it's half the system. What use is a tape without its (exactly right) player? No Jurassic Park, then.

We could perhaps invent experimental systems that "re-play" an extinct animal, that look more feasible than dino-saurs, but there must, both in principle and in practice, be enormous difficulties. It is instructive to work out just how difficult such a piece of bioengineeering would be; it would cost a lot more than the *Jurassic Park* film, even for a much easier challenge. What about mammoths, whose flesh we've actually got frozen (the DNA is less degraded than dodos')? That would be nearly as much fun as dinosaurs, and *much, much* easier. All we'd have to do is to find a hormone-injection regime that provides viable elephant eggs; work out the proper saline solutions for elephant eggs to be happy in; and fix temperature, oxygen, and CO_2 concentrations for them to develop robustly. It took about a million mouse eggs to achieve the mouse system (and we still can't replace nuclei with different DNA); about two million cattle eggs to achieve that (very different) system; and we still can't do it reliably with hamsters after about four million eggs in culture. The human test-tube-baby system is very robust, and we developed it relatively easily (a few thousand eggs), because the parameters are surprisingly close to mouse. Let's imagine the test-tube elephant works after only a million eggs. After ten years of a thousand elephants, ten experi-mental cycles a year (over)producing ten eggs each cycle, we've got our system which might occasionally accept a per-fect mammoth nucleus. (We haven't *got* any perfect mam-moth nuclei, incidently—God doesn't freeze them carefully enough.) Then we discover (if the mother elephant doesn't react against the foreign mammoth proteins of the embryo) that elephant milk doesn't work for mammoths—how many baby mammoths will we use up to discover that? Inciden-tally, they *won't* be baby mammoths; they'll be what you get when you play mammoth DNA tape in an elephant egg, then mature the products in an elephant uterus. Near-mammoths? Perhaps if you could get the near-mammoths to breed with each other, the next generation might have the right ancient ovary, and their progeny might be nearer-mammoths—but how would we know? It's probably not worth spending money as do physicists or astronomers to produce this new organism—real mammoths are *extinct.* Finis. So are dodos. The amount of effort required to reconstruct a devlopmental program is immense. Don't get the idea, which some simple-minded newspaper articles promote, that we've "conserved" an animal—or a plant—if we've got its DNA. The mammoth example shows you a few of the difficulties.

So much for the notion that DNA *determines* what an organism is like; it doesn't. There is, in principle, no one-to-one relationship, no "mapping," from DNA sequences to characters. (Of course, we can map *differences* of character—like albinism or Parkinson's disease—to specific differences in DNA.) The whole process of development, from ovary-making egg to mother-making ovary, holds itself together. Each bit of information context, like the egg mechanisms, is necessary and specific for each bit of information content, like the DNA. What makes the fly, or you, is the complete process of development. All of it. Can you blame your DNA for your funny squiggly handwriting, your passion for Fats Waller and Burmese cats, your blue eyes? Well, perhaps the last, but certainly not the others. You can't blame the DNA for what you've made of yourself. You, the process, are responsible for what you are, what you do. And for what you become.

JACK COHEN is an internationally known reproductive biologist who consults for test-tube-baby and other infertility laboratories.

A university teacher for some thirty years, he has published nearly a hundred research papers. His books include *Living Embryos,* a classic textbook whose three editions sold more than one hundred thousand copies; *Reproduction;* and *The Privileged Ape,* a rather different look at human evolution. He now works with the mathematician Ian Stewart, with whom he has explored issues of complexity, chaos, and simplicity in their first joint book, *The Collapse of Chaos.*

Cohen acts as a consultant to top science-fiction authors such as McCaffrey, Gerrold, Harrison, Niven, and Pratchett, designing alien creatures and ecologies and helping them avoid scientific blunders. He is frequently heard on BBC radio programs, and has initiated and participated in the production of several TV programs (BBC *Horizon: Genesis;* ITV series: *Take Another Look;* BBC Channel 2 series: *Fancy Fish*) for which he did much of the filming, especially time-lapse microscopy.

Triumph of the Embryo

Lewis Wolpert

H ow can something as small and dull as an egg give rise to a complex human being? Where is all the machinery to turn that little cell into all the tissues of the body? How can genes, the hereditary material, control these processes, and how can genes generate all the astonishing variety of animal life? These present one of the greatest problems in biology. And the recent progress toward answers is extremely exciting.

The fertilized egg cell gives rise to multitudes of cells— billions in humans—that become organized into structures like eyes, noses, limbs, hearts, and brains. How are the structures, or at least the plans for making them, embedded in the egg? They cannot all be preformed in the egg—development does not work simply by expanding an existing pattern—and there are clearly organizing mechanisms at work. Even when parts of the early embryo are removed, the embryo can regulate and develop normally, despite the disturbance. The development of identical human twins can result from the division of the embryo when hundreds of cells are already present.

To understand development, we must look at both cells and genes. Development is best understood in terms of cellular behavior, which is controlled by genes. Cells are the basic units in the developing embryo. The egg divides and multiplies and different types of cells arise. We can see this variety in muscle cells, nerve cells, skin cells, lens cells, and so on. We have in our bodies about two hundred and fifty different cell types. But there is more to development than just generating different cell types. The cells have to undergo pattern formation and morphogenesis—change in form. For example, they form structures like arms and legs, each of which contain very similar kinds of cells. This involves pattern formation, which gives cells a positional identity so that they can develop in an appropriate manner. Pattern formation is about spatial organization—putting muscle and bone in the right place so arms differ from legs, and bats' wings from birds' wings. Morphogenesis is concerned with the physical mechanisms whereby the embryo changes its form. For example, our brain is initially a flat sheet of cells that rolls up into the shape of a tube. This is brought about by active movement of the cells and changes in their adhesive properties that stick them together. In general, pattern formation precedes morphogenesis and tells the cells where to change shape or alter their adhesiveness.

It is difference in pattern formation and thus spatial organization rather than cell types that distinguishes us from other vertebrates. While there may be some small differences in, for example, their limb and brain cells, it is really how they are spatially organized that matters. We have no cells in our brains that chimpanzees do not have.

During development, cells multiply, change their character, exert forces, and give and receive signals. All these activities are under the control of the genetic information in the genes contained within the DNA of the chromosomes. But DNA is a rather passive and stable chemical, and the way it controls cell behavior is by controlling which proteins are made in the cell. Proteins are the magicians of the cell that make things happen. Proteins control both the chemical reaction and structures in the cell. In fact, a cell is characterized by the special proteins it has. Proteins bring about cell movement, determine the shape of the cell, and enable it to multiply. Each cell has its own set of proteins—like hemoglobin in red cells or insulin in pancreas cells, but since the DNA contains the instructions for making all proteins and each protein is coded for by a gene, whether or not a protein is present in a cell is determined by whether or not its gene is on or off. So switching genes on and off is the central feature of development, since it controls which proteins are made and the resultant cell behavior. Each cell contains the same genetic information that it gets from the egg, and differences between cells thus result from turning on and off different genes.

How do these differences in gene activity arise? Some of the differences arise because the egg is not as dull as I suggested. In frogs and flies, for example, there are special proteins in particular regions of the egg which are put there when the egg is made by the mother. So when the egg divides, some cells acquire one kind of protein and others another kind, and these proteins can activate quite different genes. But these differences in the egg only set up crude domains, and cell communication is the main method used to pattern the embryo. Human embryos seem to use intercellular communication exclusively, since there is no reason to believe that there are any differences in the egg. And, of course, the ability of embryos to regulate—that is, develop normally, when perturbed—absolutely requires communication, for how else would they make good the defect?

How then do cells in the embryo know what to do? The answer in part is that they rely on "knowing" their position. It is quite convenient to think not of embryos—which are quite complicated—but of flags. Consider the French flag problem. Imagine that there is a line of cells, each of which can develop into a blue or white or red cell. What mechanism could then reliably generate the French flag pattern—that is, the cells in the first third of the line will become blue, the next third white, and the final third red? This is not as remote as it might seem from the problem that cells in the early embryo have to face, since at an early stage, the embryo becomes divided into several regions that give rise to, for example, the skeleton and muscles, the gut and the skin.

There are a number of solutions to the problem, but probably the most general and interesting is that each cell requires positional identity. If the cells "know" their position with respect to the end of the line, then using their genetic instructions, which are identical for each cell, they could work out which third part they were in and so develop into blue, white, or red cells. One way this would work might rely on a graded concentration of a chemical—a morphogen—along the line; by reading the concentration of the morphogen, the cells would know their position. At high concentrations, red cells would develop, and so on. More generally, if cells have their positions specified and have genetic instructions as to what to do at each position, a wide variety of patterns could be generated.

A nice example of how the position of cells is signaled is illustrated by a classic experiment with early frog development. The embryonic pattern is initially specified within the surface layers of the spherical embryo that has come from the division of the egg. This gives essentially a two-dimensional pattern and regions like those that will form the gut and skeleton are still on the outside of the embryo. They now move inside during a process known as gastrulation. The site where they enter the embryo and move inside is a signaling region responsible for establishing the basic pattern of the main body axis. If this region is grafted to another embryo, it gives a signal to his host so that a whole new embryo can be induced, both head and body.

Another example of positional signaling is seen in limb development. The signaling region is at the posterior margin of the bud, and gives the digits their positional identity; if this region is grafted to the anterior margin of another bud, its signal results in a mirror-image limb developing with two sets of digits. One way of thinking of this is in terms of the signal establishing a mirror-image gradient in a morphogen.

The cells also need to record and remember their positional identity. Studies on early insect development have been wonderfully successful in identifying the genes responsible for patterning the early embryo. It turns out that the identity of the different parts of the insect body are controlled by genes which have a particular character and are called homeotic genes. Mutations in these genes can convert one part of the body into another like an antenna of a fly into a leg, a process known as homeosis. These genes have one small region in common which is known as the homeobox. Remarkably, homeobox-containing genes are also present in all other animals, where they seem to serve similar functions, in recording positional identity. There is a well-defined pattern of homeobox gene expression along the main body axis among mouse and frog embryos, probably giving the cells their positional identity. If these genes aren't expressed in the right place, then ribs, for example, may also develop in the wrong place. Similarly, in the limb homeobox, genes are present in a well-defined pattern. Moreover, when one makes a mirror-image limb by grafting a new signaling region, an early response is a change in the expression of the homeobox genes. But we still have a lot of detail to work out to understand how to get, for example, from signals and homeobox genes to the five fingers of the hand with its complex arrangements of muscles, bones, and tendons.

However different a fly may look compared to a mouse or a human being, what recent advances in molecular embryology has shown us is that they develop using very similar mechanisms. They even use very similar genes. There is now evidence that the genes and signals patterning the fly wing and vertebrate limb are similar. It is subtle differences in genes that modify cell behavior during development, and so generate the diversity of animal life.

LEWIS WOLPERT holds the post of professor of biology as applied to medicine at University College and Middlesex School of Medicine. His research interests are in cell and developmental biology.

In 1968, he was awarded the Scientific Medal of the Zoological Society. He was made a fellow of the Royal Society in 1980, and is chairman of COPUS (Committee for the Public Understanding of Science). He was awarded the C.B.E (Companion of the Order of the British Empire) in 1990. He is also chairman of the Medical Research Council Committee on the Genetic Approach to Human Health.

In 1986, he gave the Royal Institution Christmas Lectures, and in the spring of 1990, the Radcliffe Lectures at the University of Warwick. He was presenter for the television science program

Antenna (BBC2) in 1988–89, and has done twenty-five interviews with scientists on Radio 3. The first radio series was published in *A Passion for Science.* He has also made a number of documentaries, including *The Dark Lady of DNA* and *The Virgin Fathers of the Calculus.* He is author of *The Triumph of the Embryo* and *The Unnatural Nature of Science.* His latest book, *Is Science Dangerous?* appears in the series *Contemporary Papers* by W. H. Smith.

From Kefir to Death

Lynn Margulis

〰〰〰〰〰〰 It happens to the "individual." Death is the arrest of the self-maintaining processes we call metabolism, the cessation, in a given being, of the incessant chemical reassurance of life. Death, signaling the disintegration and dispersal of the former individual, was not present at the origin of life. Unlike humans, not all organisms age and die at the end of a regular interval. The aging and dying process itself evolved, and we now have an inkling of when and where. Aging and dying first appeared in certain of our microbial ancestors, small swimmers, members of a huge group called "protoctists." Some two billion years ago, these ancestors evolved both sex by fertilization and death on cue. Not animals, not plants, not even fungi nor bacteria, protoctists form a diverse—if obscure—group of aquatic beings, most of which can be seen only through a microscope. Familiar protoctists include amoebae, euglenas, ciliates, diatoms, red seaweeds and all other algae, slime molds, and water molds. Unfamiliar protoctists have strange names: foraminifera, heliozoa, ellobiopsids, and xenophyophores. An estimated two hundred and fifty thousand species exist, most of which have been studied hardly at all.

Death is the loss of the individual's clear boundaries; in death, the self dissolves. But life in a different form goes on—as the fungi and bacteria of decay, or as a child or a grandchild who continues living. The self becomes moribund because of the disintegration of its metabolic processes, but metabolism itself is not lost. Any organism ceases to exist because of circumstances beyond its control: the ambience becomes too hot, too cold, or too dry for too long; a vicious predator attacks or poison gas abounds; food disappears or starvation sets in. The causes of death in photosynthetic bacteria, algae, and plants include too little light, lack of nitrogen, or scarcity of phosphorus. But death also occurs in fine weather independently of direct environmental action. This built-in death—for example, Indian corn stalks that die at the end of the season and healthy elephants that succumb at the end of a century—is programmed. Programmed death is the process by which microscopic protoctists—such as *Plasmodium* (the malarial parasite) or a slime mold mass—dry up and die. Death happens as, say, a butterfly or a lily flower made of many cells matures and then disintegrates in the normal course of development.

Programmed death occurs on many levels. Monthly, the uterine lining of menstruating women sheds as its dead cells (the menstrual blood) flow through the vagina. Each autumn, in deciduous trees and shrubs of the north temperate zone, rows of cells at the base of each leaf stem die. Without the death of this thin layer, cued by the shortening of day length, no leaf would fall. Using genetic-engineering techniques, investigators such as my colleague at the University of Massachusetts, Professor Lawrence Schwartz, can put certain "death genes" into laboratory-grown cells which are not programmed to die. The flaskful of potentially immortal cells, on receipt of this DNA, then die so suddenly that the precipitous cessation of their metabolism can be timed to the hour. The control cells that have not received the death genes live indefinitely. Menstrual blood, the dying leaf layer, the rapid self-destruction of the cells that receive the "death genes," and the slower, but more frightening aging of our parents and ourselves are all examples of programmed death.

Unlike animals and plants that grow from embryos and die on schedule, all bacteria, most nucleated microscopic beings, namely the smaller protoctists and fungi such as molds and yeast, remain eternally young. These inhabitants of the microcosm grow and reproduce without any need for sexual partners. At some point in evolution, meiotic sex—the kind of sex involving genders and fertilization—became correlated with an absolute requirement for programmed death. How did death evolve in these protoctist ancestors?

An elderly man may fertilize a middle-aged woman, but their child is always young. Sperm and egg merge to form the embryo which becomes the fetus and then the infant. Whether or not the mother is thirteen or forty-three years old, the newborn infant begins life equally young. Programmed death happens to a body and its cells. By contrast, the renewed life of the embryo is the escape from this predictable kind of dying. Each generation restores the *status quo ante,* the microbial form of our ancestors. By a circuitous route, partners that fused survived, whereas those that never entered sexual liaisons passed away.

Eventually, the ancestral microbes made germ cells that frantically sought and found each other. Fusing, they restored youth. All animals, including people, engage in meiotic sex; all descended from microbes that underwent meiosis (cell divisions that reduce by half chromosome numbers) and sex (fertilization which doubles chromosome numbers).

Bacteria, fungi, and even many protoctists were—and are—reproducing individuals that lack sex lives like ours. They must reproduce without partners, but they never die unless they are killed. The inevitability of cell death and the mortality of the body is the price certain of our protoctist ancestors paid—and we pay still—for the meiotic sex they lack.

Surprisingly, a nutritious and effervescent drink called kefir, popular in the Caucausus Mountains of southern Russia and Georgia, informs us about death. Even more remarkably, kefir also illustrates how symbiogenesis—the appearance of new species by symbiois—works. The word *kefir* (also spelled *kephyr*) applies both to the dairy drink and to the individual curds or grains that ferment milk to make the drink. These grains, like our protoctist ancestors, evolved by symbiosis.

Abe Gomel, the Canadian businessman and owner of Liberté (Liberty) dairy products, manufactures real kefir of the Georgian Caucausus as a small part of his line of products. He and his diligent coworker Ginette Beauchemin descend daily to the basement vat room of his factory to inspect the heated growth of the thick, milky substance on its way to becoming commercial kefir. Like all good kefir makers, they know to transfer the most plump and thriving pellets between nine and ten every morning, weekends included, into the freshest milk. Although nearly everyone who lives in Russia, Poland, or even Scandinavia drinks kefir, this "champagne yoghurt" of the Caucasian peoples is still almost unknown in western Europe and the Americas. Abe Gomel and Ginette Beauchemin have been able to train only two other helpers, who must keep constant vigil over the two vats that are always running.

Legend says the prophet Muhammad gave the original kefir pellets to the Orthodox Christian peoples in the Caucausus, Georgia, near Mount Ebrus, with strict orders never to give them away. Nonetheless, secrets of preparation of the possibly life-extending "Muhammad pellets" have of course been shared. A growing kefir curd is an irregular spherical being. Looking like a large curd of cottage cheese, about a centimeter in diameter, individual kefir pellets grow and metabolize milk sugars and proteins to make kefir the dairy drink. When active metabolism that assures individuality ceases, kefir curds dissolve and die without aging. Just as corncobs in a field, active yeast in fermenting vats, or fish eggs in trout hatcheries must be tended, so kefir requires care. Dead corn seeds grow no stalks, dead yeast makes neither bread nor beer, dead fish are not marketable, and in the same way, kefir individuals after dying are not kefir. Comparable to damp but "inactive" yeast or decaying trout eggs, dead kefir curds teem with a kind of life that is something other than kefir: a smelly mush of irrelevant fungi and bacteria thriving and metabolizing, but no longer in integrated fashion, on corpses of what once were live individuals.

Like our protoctist ancestors that evolved from symbioses among bacteria, kefir individuals evolved from the living together of some thirty different microbes, at least eleven of which are known from recent studies (see table, page 74). These specific yeasts and bacteria must reproduce together— by coordinated cell division that never involves fertilization or any other aspect of meiotic sex—to maintain the integrity of the unusual microbial individual that is the kefir curd. Symbiogenesis led to complex individuals that die (like kefir and most protoctists) before sexuality led to organisms that *had* to die (like elephants and us). A kefir individual, like any other, requires behavioral and metabolic reaffirmation.

KEFIR: List of Components, Live Microbes

Each individual (see figure) is composed of:

Kingdom Monera (bacteria)
  Streptococcus lactis
  Lactobacillus casei, Lactobacillus brevis
  Lactobacillus helveticus, Lactobacillus bulgaricus
  Leuconostoc mesenteroides
  Acetobacter aceti

Kingdom Fungi (yeasts, molds)
  Kluyveromyces marxianus, Torulaspora delbrueckii
  Candida kefir, Saccharomyces cerevisiae
  and at least 15 other kinds of unknown microbes

During the course of brewing the yogurtlike beverage, people inadvertently bred for kefir individuals. In choosing the best "starter" to make the drink, villagers of the Caucasus "naturally selected," which means they encouraged the growth of certain populations and stopped the growth of others. These people inadvertently turned a loose confederation of microbes into well-formed populations of much larger individuals, each capable of death. In trying to satisfy their taste buds and stomachs, kefir-drinking Georgians are unaware that they have created a new form of life.

The minute beings making up live kefir grains can be seen with high-power microscopy (see figure): specific bacteria and fungi inextricably connected by complex materials, glycoproteins and carbohydrates of their own making—individuals bounded by their own skin—so to speak. In healthy

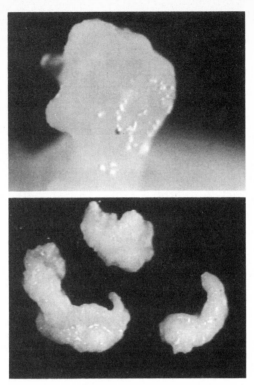

Figure: A kefir grain: the complex kefir "individual" live as seen by low-power microscopy (magnification 5 times). Top—one individual; bottom—three individuals

kefir, the bacterial and fungal components are organized into a curd, a covered structure which reproduces as a single entity. As one curd divides to make two, two become four, eight, sixteen, and so on. The reproducing kefir forms the liquid that after a week or so of growth becomes the dairy drink. If the relative quantities of its component microbes are skewed, the individual curd dies and sour mush results.

Kefir microbes are entirely integrated into the new being just as the former symbiotic bacteria which became components of protoctist and animal cells are integrated. As they grow, kefir curds convert milk to the effervescent drink. "Starter," the original Caucasian kefir curds, must be properly tended. Kefir can no more be made by the "right mix" of chemicals or microbes than can oak trees or elephants.

Scientists now know, or at least strongly suspect from DNA sequence and other studies, that the oxygen-using parts of nucleated cells evolved from symbioses when certain fermenting larger microbes (thermoplasma-like archaebacteria) teamed up with smaller, oxygen-respiring bacteria.

Mitochondria, which combine oxygen with sugars and other food compounds to generate energy, are found almost universally in the cells of protoctists, fungi, plants, and animals. We, as all mammals, inherit our mitochondria from our mother's egg. Like kefir, we, and all other organisms made of nucleated cells, from amoebae to whales, are not only individuals, we are aggregates. Individuality arises from aggregation, communities whose members fuse and become bounded by materials of their own making. Just as people unconsciously selected the new kefir life-form, so other beings caused evolution of new life—including our ancestors—as microbes, feeding on each other's fats, proteins, carbohydrates, and waste products but only incompletely digesting them, selected each other and eventually coalesced.

Plants come from ancestors that selected but did not entirely digest each other as food. Hungry ancestral swimming cells took up green photosynthetic microbes called cyanobacteria. Some resisted digestion, surviving inside larger cells and continuing to photosynthesize. With integration, green food grew as a part of a new self. The bacterium outside was now an integral part inside the cell. From partly digested cyanobacterium and a hungry translucent swimmer, a new individual, the alga, evolved. From green algal cells (protoctists) came the cells of plants.

Kefir is a sparkling demonstration that the integration processes by which our cells evolved still occurs. Kefir also helps us recognize how the origin of a complex new individual preceded programmed death of the individual on an evolutionary time scale. Kefir instructs us, by its very existence, about how the tastes and choices of one species (ours) influence the evolution of others, the thirty intertwined microbes that became kefir. Although kefir is a complex individual, a product of interacting aggregates of both nonnucleated bacteria and nucleated fungi, it reproduces by direct growth and division. Sex has not evolved in it, and, relative to elephants and corn stalks, both of which develop from sexually produced embryos, kefir grains undergo very little development and display no meiotic sexuality. Yet when mistreated they die and, once dead, like any live individual, never return to life as that same individual.

Knowing that symbionts become new organisms illuminates individuality and death. Individuation, which evolved in the earliest protoctists in a manner similar to the way it did in kefir, preceded meiotic sexuality. Programmed aging and death was a profound evolutionary innovation limited to the descendants of the sexual protoctists that became animals, fungi, and plants.

The development of death on schedule, the first of the sexually transmitted diseases, evolved along with our peculiar form of sexuality, a process that kefir lacks now and always has done without. The privilege of sexual fusion— the two-parent "fertilization-meiosis" cycle of many protoctists, most fungi, and all plants and animals—is penalized by the imperative of death. Kefir, by not having evolved sex, avoids having to die by programmed death.

LYNN MARGULIS is a biologist, and Distinguished University Professor in the Department of Biology at the University of Massachusetts at Amherst. She held a Sherman Fairchild Fellowship at the California Institute of Technology (1977) and a Guggenheim Fellowship (1979). She is a member of the National Academy of Sciences. From 1977 to 1980, she chaired the National Academy of Science's Space Science Board Committee on Planetary Biology and Chemical Evolution, aiding in the development of research strategies for NASA for which she received a NASA public service award in 1981.

Her published writings, spanning a wide range of scientific topics, range from professional to children's literature and include: *The Origin of Eukaryotic Cells* (1970); *Early Life* (1981); and *Symbiosis in Cell Evolution* (2d ed., 1993). She is coauthor, with Karlene V. Schwartz, of *Five Kingdoms: An Illustrated Guide to the Phyla of Life on Earth* (2d ed., 1988); and with Dorion Sagan, *Microcosmos* (1986); *Origins of Sex* (1986); *Mystery Dance* (1991); and *What Is Life?* (1995). She has participated in the development of science teaching materials at levels from elementary through graduate school. She collaborates with James E. Lovelock, F.R.S., on investigations concerning his "Gaia Hypothesis." She also contributes to research on cell biology and on microbial evolution.

Part Three

Evolution

Three Facets of Evolution

Stephen Jay Gould

1. What Evolution Is Not

Of all the fundamental concepts in the life sciences, evolution is both the most important and the most widely misunderstood. Since we often grasp a subject best by recognizing what it isn't, and what it cannot do, we should begin with some disclaimers, acknowledging for science what G. K. Chesterton considered so important for the humanities: "Art is limitation; the essence of every picture is the frame."

First, neither evolution, nor any science, can access the subject of ultimate origins or ethical meanings. (Science, as an enterprise, tries to discover and explain the phenomena and regularities of the empirical world, under the assumption that natural laws are uniform in space and time. This restriction places an endless world of fascination within the "picture"; most subjects thus relegated to the "frame" are unanswerable in any case.) Thus, evolution is not the study of life's ultimate origin in the universe or of life's intrinsic significance among nature's objects; these questions are philosophical (or theological) and do not fall within the purview of science. (I also suspect that they have no universally satisfactory answers, but this is another subject for another time.) This point is important because zealous fundamentalists, masquerading as "scientific creationists," claim that creation must be equated with evolution, and be given equal time in schools, because both are equally "religious" in dealing with ultimate unknowns. In fact, evolution does not treat such subjects at all, and thus remains fully scientific.

Second, evolution has been saddled with a suite of concepts and meanings that represent long-standing Western social prejudices and psychological hopes, rather than any account of nature's factuality. Such "baggage" may be unavoidable for any field so closely allied with such deep human concerns (see part 3 of this statement), but this strong social overlay has prevented us from truly completing Darwin's revolution. Most pernicious and constraining among these prejudices is the concept of progress, the idea that evolution possesses a driving force or manifests an overarching trend toward increasing complexity, better biomechanical design, bigger brains, or some other parochial definition of progress centered upon a long-standing human desire to place ourselves atop nature's pile—and thereby assert a natural right to rule and exploit our planet.

Evolution, in Darwin's formulation, is adaptation to changing local environments, not universal "progress." A lineage of elephants that evolves a heavier coating of hair to become a woolly mammoth as the ice sheets advance does not become a superior elephant in any general sense, but just an elephant better adapted to local conditions of increasing cold. For every species that does become more complex as an adaptation to its own environment, look for parasites (often several species) living within its body—for parasites are usually greatly simplified in anatomy compared with their free-living ancestors, yet these parasites are as well adapted to the internal environment of their host as the host has evolved to match the needs of its external environment.

2. What Evolution Is

In its minimalist, "bare bones" formulation, evolution is a simple idea with a remarkable range of implications. The basic claim includes two linked statements that provide rationales for the two central disciplines of natural history: taxonomy (or the order of relationships among organisms), and paleontology (or the history of life). Evolution means (1) that all organisms are related by ties of genealogy or descent from common ancestry along the branching patterns of life's tree, and (2) that lineages alter their form and diversity through time by a natural process of change—"descent with modification" in Darwin's chosen phrase. This simple, yet profound, insight immediately answers the great biological question of the ages: What is the basis for the "natural system" of relationships among organisms (cats closer to dogs than to lizards; all vertebrates closer to each other than any to an insect—a fact well appreciated, and regarded as both wonderful and mysterious, long before evolution provided the reason). Previous explanations were unsatisfactory because they were either untestable (God's creative hand making each species by fiat, with taxonomic relationships representing the order of divine thought), or arcane and complex (species as natural places, like chemical elements in the periodic table, for the arrangement of organic matter). Evolution's explanation for the natural system is so stunningly simple: Relationship is genealogy; humans are like apes because we share such a recent common ancestor. The taxonomic order is a record of history.

But the basic fact of genealogy and change—descent with modification—is not enough to characterize evolution as a science. For science has two missions: (1) to record and discover the factual state of the empirical world, and (2) to devise and test explanations for why the world works as it does. Genealogy and change only represent the solution to this first goal—a description of the fact of evolution. We also need to know the mechanisms by which evolutionary change occurs—the second goal of explaining the causes of descent with modification. Darwin proposed the most famous and best-documented mechanism for change in the principle that he named "natural selection."

The fact of evolution is as well documented as anything we know in science—as secure as our conviction that Earth revolves about the sun, and not vice versa. The mechanism of evolution remains a subject of exciting controversy—and science is most lively and fruitful when engaged in fundamental debates about the causes of well-documented facts. Darwin's natural selection has been affirmed, in studies both copious and elegant, as a powerful mechanism, particularly in evolving the adaptations of organisms to their local environments—what Darwin called "that perfection of structure and coadaptation which most justly excites our admiration." But the broad-scale history of life includes other phenomena that may require different kinds of causes as well (potentially random effects, for example, in another fundamental determinant of life's pattern—which groups live, and which die, in episodes of catastrophic extinction).

3. Why Should We Care?

The deepest, in-the-gut, answer to this question lies in the human psyche, and for reasons that I cannot begin to fathom. We are fascinated by physical ties of ancestry; we feel that we will understand ourselves better, know who we are in some fundamental sense, when we trace the sources of our descent. We haunt graveyards and parish records; we pore over family Bibles and search out elderly relatives, all to fill in the blanks on our family tree. Evolution is this same phenomenon on a much more inclusive scale—roots writ large. Evolution is the family tree of our races, species, and lineages—not just of our little, local surname. Evolution answers, insofar as science can address such questions at all, the troubling and fascinating issues of "Who are we?" "To which other creatures are we related, and how?" "What is the history of our interdependency with the natural world?" "Why are we here at all?" Beyond this, I think that the importance of evolution in human thought is best captured in a famous statement by Sigmund Freud, who observed, with wry and telling irony, that all great scientific revolutions have but one feature in common: the casting of human arrogance off one pedestal after another of previous convictions about our ruling centrality in the universe. Freud mentions three such revolutions: the Copernican, for moving our home from center stage in a small universe to a tiny peripheral hunk of rock amid inconceivable vastness; the Darwinian, for "relegating us to descent from an animal world"; and (in one of the least modest statements of intellectual history) his own, for discovering the unconscious and illustrating the nonrationality of the human mind. What can be more humbling, and therefore more liberating, than a transition from viewing ourselves as "just a little lower than the angels," the created rulers of nature, made in God's image to shape and subdue the earth—to the knowledge that we are not only natural products of a universal process of descent with modification (and thus kin to all other creatures), but also a small, late-blooming, and ultimately transient twig on the copiously arborescent tree of life, and not the foreordained summit of a ladder of progress. Shake complacent certainty, and kindle the fires of intellect.

STEPHEN JAY GOULD is an evolutionary biologist, a paleontologist, and a snail geneticist; professor of zoology, Harvard University; MacArthur fellow; and author of numerous books, including *Ontogeny and Phylogeny; The Mismeasure of Man; The Flamingo's Smile; Wonderful Life;* and *Bully for Brontosaurus.* He is internationally recognized for his scientific contributions in paleontology and evolutionary biology and also for his capacity to communicate ideas to the general public.

Our Gang

Milford H. Wolpoff

Perhaps more than most, paleoanthropology is known as the science notorious for the importance of new discoveries. The newspapers love it, as time after time a new fossil is dramatically announced that is said to catch scientists by surprise, or disprove conclusions. But if paleoanthropologists were constantly being thrown off guard, their theories couldn't be very good. I once got so irritated by these announcements, and the poor way they reflected on the ability of paleoanthropology to develop robust theories about human evolution, that in a fit of dark humor I proposed to present a paper at the national meetings of the biological anthropologists entitled "Phenomenal New Discovery Overturns All Previously Held Theories of Human Evolution." In what was probably a fit of even darker humor, the referee committee accepted the title, forcing me to write the paper.

Science needs a body of knowledge to explain, and a science seems to progress as new discoveries create the need to alter earlier explanations. Yet, the real advances we make in understanding our past come from the discoveries of the mind, the unique ideas that new thinking brings to the way we organize the world, because the way we organize it becomes the way we understand it. Our mental constructs— our theories—are based on facts because they have to explain them; yet they have an independence from reality, a life of their own. By outlining the basis of interrelationships, theories go well beyond the facts, and eventually become so convincing in their organization and explanation of data that they are difficult to dismiss even in the face of conflicting information. Whether it is called insight, sideways thinking, or whatever, theory building is where the real progress lies.

Evolution, the recognition that all living forms are related through common descent and that genetic changes lead to the diversity of life on earth, is one of these revolutionary advances in how we organize our knowledge. It is arguably *the* revolutionary advance in biology. Yet despite the fact that evolution is an explanation of *our* ancestry and *our* diversity, or perhaps because of the fact, evolution is widely disbelieved. Most distasteful to the skeptical are the implications that descent with modification has for understanding the true relationship we have to other animals—it seems to undermine what everybody knows: that we humans are unique and special. Most difficult to swallow is what every schoolchild learns, that "man is descended from the apes." Is this really true?

Watching *Inherit the Wind* recently, the film version of the 1925 Scopes trial, I was struck by its rendition of the carnival atmosphere in the town, just outside the courthouse where the trial was about to take place. There is a musical background of *Give Me That Old-Time Religion,* and a visual background of waving placards and posters proclaming that "man did *not* evolve from the apes." Then, as if the height of ridicule, a carnival exhibit features a cigarette-smoking chimpanzee sitting under a sign that offers the message, DARWIN WAS WRONG—THE APES EVOLVED FROM MAN! Now, generations after the trial and long after the film was made, I have come to wonder how ridiculous this message actually is? Now the first one, but the second. Just what *is* our relation to apes?

In part, my probing and questioning of what would seem to be a parody of evolutionary science, in a movie scene meant to show the strength of antievolutionary feeling, is a direct consequence of today's vastly increased knowledge of our past. Simply put, we see things more clearly than when Scopes challenged the state of Tennessee, although I must admit that our vision is still, by any standard, very blurred at best.

To understand what it is like to peer into the past with the frustrations of an evolutionary scientist, imagine setting up two mirrors facing each other. If you step between them, positioning yourself just right, you can see yourself in the first mirror exactly as you stand before it. You will also see a second, smaller image of yourself reflected from the second mirror into the first mirror, and an even smaller image reflected from the first mirror to the second and then back to the first, and so on. As long as you are imagining, mirrors are not perfect, and suppose that a close examination of the images shows that the first, biggest one looks slightly different from you. The next smaller one is more different, and the next smaller one is more different yet. Just when there are enough differences to make the images interesting, they are so small that you can't see them clearly anymore. Those out-of-focus images are the evidence we have for what changed, the basis for evolutionary science. And every now and then, there is a particularly clear image, a "new discovery." Yet, as important as they are, the images are not of evolution, but of the domain that evolution explains. What progress there is in understanding human evolution could never result from this tunnel view into the past alone. Our expectations are far more important than the discoveries that might confirm or reject them. I am not the first to note that we do not see things as they are, but as we are.

Probing into our past raises some philosophical issues, because it involves the most fundamental questioning of humanity's place in nature. It also raises an issue of taxonomy, the theory and practice of classifying organisms according to their closeness of relationship. Taxonomy is meant to reflect the evolutionary process—the most similar-appearing species are usually assumed to be the most closely related. But in this century, we have learned that similarities in appearance may be a misleading guide to relationship, and in the process of agonizing over which are the most important characteristics for classification, it has been easy to lose sight of the fact that taxonomy is not a classification of similarities but a genealogy of species. Taxonomic names and their arrangement in a family tree of species reflect historic relationships and not similarities, just as the terms of human kinship (father, cousin) reflect the genealogy of individuals and not how much they look alike. But like any spin-off of scientific thinking (evolutionary history is, after all, a scientific hypothesis about what we think happened in the past), taxonomy probably reflects the prejudices and preconceptions of taxonomists to a significant degree as well. And when the object of taxonomic focus is humanity, taxonomy becomes a tool for establishing our important position and dominance, setting humanity apart and establishing a kind of them versus us. In fact, our assessment of the importance and uniqueness of humankind has stood in the way of understanding our place in nature and our true relation to the other species on the planet.

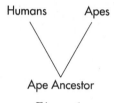

Figure 1
The Classic Darwinian View of Human Descent

"Man and the apes" is an expression of a taxonomic relationship, as well as an unfortunate turn of phrase that is not only sexist, but now appears to be overly anthropocentric as well. Great apes—the gorilla and chimpanzee of Africa and the orangutan and recently extinct *Gigantopithecus* of south and southeast Asia—have long been held as the evolutionary contrast to humans, visualized almost as failed attempts to become human when they are described in terms of which human features they show, and which they lack. The claim that "man descended from the apes" contains an implicit notion of our poor opinion of these unfortunate primates who never evolved. Realizing that humans could not have descended from *living* apes, this phrase can only mean that these large primates are similar to our common ancestor (Figure 1). In this meaning there is the assumption that while human evolution continued to its present heights, evolution stopped for the apes before it ever had a chance to get started. We are the featherless bipeds, with our large brains set above furless bodies and our hands freed from the requirements of locomotion to carry and use tools and manipulate the environment. The other, in contrast, are the bestial, hairy, long-armed tree dwellers of Tarzan fame, who never fully made the grade. The taxonomists reflect this in their classification of the large primates: the hominids and the pongids (or if you prefer the formality and the Latin, *Hominidae* and *Pongidae*—two families in the classification of the anthropoid primates). The hominids are the humans and all of our ancestors back to the split with apes, and the pongids are the apes and all of their ancestors back to the split with humans. Thus, by this obvious classification, we are formally set apart from them—and why not? Aren't we *really* different?

Everyone knows there is a gap between humans and apes, a series of consistent differences that helps outline our humanity. Some of these differences are in features unique to us—for instance: bipedalism (upright posture, two-legged locomotion); large brains and the very complex behaviors they allow (social interactions, language, stratified roles, etc.); weak and relatively hairless bodies that make humans rely on knowledge and technology; skill and cunning, to survive and persist. Other differences are reflected in the commonalities uniquely shared by the apes, and not by humans. These include similarities of habitat (many live in dense tropical forest); similarities of size and form (for example, very elongated forelimbs relative to the hindlimbs, a short lumbar spine); a strange mix of arboreal movements (climbing, hanging by the arms, grasping with the feet, and a limited amount of the arm-over-arm arboreal locomotion called brachiation); and a terrestrial adaptation that shows these same primates to be capable of both bipedal walking and walking on all four limbs with their knuckles or fists rather than the palms of the forelimbs, on the ground.

As the taxonomy seems merely to state these facts in a formal, systematic way, it would appear to be unquestionable. But there is a problem. The taxonomic contrast between hominids and pongids is between closely related species that we call sister groups. This term means exactly what it sounds like: Sister species are closely related groups that share an immediate common parentage. You and your sibling are an example of a sister group made up of individuals. However, you and your cousin are *not* a sister group, because your sib shares a more recent parentage with you than your cousin does.

The fact that the common parent of "man and ape" was neither a human nor an ape has been realized for a long time, although the popular accounts still have us descending from apes (this is more often used in evolution bashing than in evolution teaching). However, that is not the problem. The problem comes from the last decade's research that shows that humans *are not* the sister group of apes! A good deal of this research is paleontological. In particular, it hinges on the analysis and interpretation of a ten- to fourteen-million-year-old fossil primate called *Ramapithecus*—a primate once thought to be a hominid because it was considered a unique human ancestor.

The relationship of *Ramapithecus* to the human line (Figure 2) was first determined from a study of the characteristics of fragments of jaws and teeth that were discovered in the Siwalik hills of the Indian subcontinent, just a few years after the Scopes trial. When the first fossilized bones and teeth of *Ramapithecus* were found, they were thought to be the remains of a human ancestor because they fit Darwin's theory of human origins.

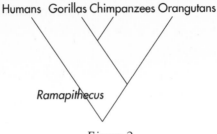

Figure 2
The Conventional Classification

Darwin addressed what he believed were the four unique characteristics of humanity: bipedal locomotion, tool use, reduction of the canine teeth, and increase in brain size. He maintained that the first humans were the result of a terrestrial adaptation and a dietary change from leaf and fruit eating to hunting. In his chain of reasoning, tool use became important for obtaining meat (this is when the canines reduced); bipedalism freed the hands to carry and use tools and weapons; and the much more complex learning that technology required (and the need for social control that it created) led to the changes in the brain.

The first *Ramapithecus* remains seemed to have the small canine teeth that Darwin's model regarded as the consequence of replacing the cutting function of the canines with hand-held tools at the time of human origins. This was part of a positive feedback relation between the four unique human characteristics that Darwin envisaged. As more *Ramapithecus* remains were slowly recovered, attempts were made to show how they continued to fit into this Darwinian model, even to the extent that a search began for evidence of the stone tools that these earliest hominids "must have" been making. Long before any limb bones of *Ramapithecus* were found, it was proclaimed to be a biped. Such is the power of theory over evidence that the fragmentary remains recovered for *Ramapithecus* were "known" to be those of the earliest hominid because they fit Darwin's theory of hominid origins.

Another importance of *Ramapithecus* at the time was that it indicated that the human line was ancient. Human ancestors were present in the middle of the Miocene period, at least fourteen million years ago, long before the living ape species diverged from each other. It seemed to show what comparative anatomists had postulated, that humans are the sister group of all apes. What changed this precept were two discoveries—one of some fossils and the other in a genetics laboratory—and a reminder that taxonomy is meant to show genealogy and not necessarily similarity.

The problem surfaced when more complete remains of *Ramapithecus* were found. Better-preserved specimens showed that the earlier interpretations were mistaken and that the true relationship of this ancient Asian primate is not with humans but with Asian apes, like the living orangutans. The similarities to humans were still real, but when they were seen in the context of the more complete specimens, it became evident to scientists that the similar characteristics reflected the anatomy of the common ancestor for all apes and humans, and were not features uniquely shared by *Ramapithecus* and humans because of some special relationship between them. The issue might be thought of in terms of who diverged first. The earlier thinking was that humans diverged first and thereby became a sister group of all Asian and African ape species. The more complete fossil *Ramapithecus* remains show that the orangutans diverged first (Figure 3), and that they are the sister group of the African apes (chimpanzees and gorillas) *and humans.*

Figure 3
The Current Classification

Additional evidence for the relationship between apes and humans is based on a quite different source—genetic comparisons of the living species. At first, several decades ago, the genetic relationships were estimated from indirect studies of gene products such as proteins. That research indicated orangutans were equally related to the African apes and humans; that is, they diverged first. This divergence was at about the time of *Ramapithecus,* so it could not be a hominid. Because these results conflicted with the accepted hominid interpretation of *Ramapithecus,* many paleoanthropologists placed their bets with the source of data they knew best, continuing to support the hominid interpretation because they believed the fossil remains showed it. Genetic anthropologists, of course, placed their bets with the genetics that they knew best, and a long-lasting enmity began. Nowadays, it is routine to directly compare DNA structures themselves (i.e., the exact sequence of base pairs on a DNA molecule) to ascertain the order of divergence. With reasonable certainty, we can now determine which two species of a three-species group diverged most recently (i.e., which two are sister species).

The surprise is that the sister species for humans is not the group of great apes (Figure 1) as the taxonomy suggests, and not even the group of African apes (chimpanzee and gorilla) as the genetic data first indicated, but is almost certainly the chimpanzee alone (Figure 3). For the sister group of humans and chimpanzees, the closest relative is the gorilla. The large Asian ape, the orangutan, is a sister group to all three of these African primates, a fact in concordance with the paleontological interpretations.

Thus, the genetic evidence shows that the orangutan diverged from the African line first. This leaves a cluster of species made up of African apes and humans. To reflect the relationship, we need a special name for this African group; I have suggested *anthropithecine* from the Latin subfamily *Anthropithecinae.* Later, among the African species, the gorilla diverged from the chimpanzee/human line, and most recently, the chimpanzees and humans diverged from each other.

Geneticists and paleoanthropologists have a history of conflicting interpretations of evolution, conflicts that have often devolved to assertions that "my evidence is better than your evidence." But the proper relation of these sciences comes about when each is used as an independent test of hypotheses generated by the other. Considering genetics and paleoanthropology in this relationship, there are some clear predictions that the genetic relationships between the living species make of the African fossil record; for instance, there is an expectation that gorilla anatomy should resemble the ancestral condition for the African species, and an estimation of recent human origins (e.g., the time of human-chimpanzee divergence). How recent? Five to seven million years ago is a currently accepted estimate from mitochondrial DNA. My reading of the fossil record is that these predictions can be confirmed.

What does this mean to us? These details might seem to

make little difference in the broader question of how (and why) humans relate to the apes. But in taxonomic practice, groups classified together should each be like families lining up for their Christmas pictures—nobody is supposed to be left out. Taxonomic groups, whether they are made up of the members of your family or the group of species all of humankind belongs to, must be monophyletic, consisting of an ancestor and all its descendants. Should someone be left out of the group, it is no longer monophyletic, because it would no longer consist of *all* the descendants of an ancestor (parent, grandparent, or whatever). Leaving a few of the family's bad sheep out may make for a more acceptable picture, but taxonomists don't like it, because the group is no longer a natural one. For the human family, it is easy to make up a monophyletic group—simply include all living humans. So much for ourselves, but who else is with us in our gang?

We know who: apes. But when we examine that term in the light of the evolutionary relationships we now understand, it does not refer to a monophyletic group anymore. There is no family made up just of apes, no valid term in a classificatory scheme that refers to them, and them alone. The requirement that nobody in the family can be left out *means that humans must be left in!* Otherwise, it would be like taking a picture of your family with your brothers and sisters, parents and grandparents, aunts and uncles, and all your cousins, but without you. Our place in nature, and the true relation of humans and apes is now revealed: For apes to be a monophyletic group, humans must be included within them. If there is no taxonomic group that can include the apes without also including us, it means we are apes, by any meaning of the word.

In this altered perspective of who is in our gang, there is

some insight to be gained about ourselves. Once popular phrases like "the naked ape" were coined because of the descent-from-the-apes perception of human evolution, and widely rejected because the perception was incorrect. Descent from the apes was wrong because it was the wrong way of looking at our relation to the apes. The fact is that in terms of genealogical relationships *we are apes,* albeit of a very special kind. Human origins did not make us what we are, and it is not our genealogy that makes us unique. Our beginnings sent us down a path that has led to an extraordinary success, unknown on this planet before us and as far as we can tell, without counterparts in the universe around us. If there is something for our species to learn about itself from our blurred prehistory, it is to be found in the steps we have taken down this pathway, and not in where the path began.

MILFORD H. WOLPOFF, professor of anthropology and adjunct associate research scientist, Museum of Anthropology at the University of Michigan in Ann Arbor, is best known for his role as an outspoken critic of the "Eve" theory, which holds that modern humans are a new species that arose in Africa alone between one hundred thousand and two hundred thousand years ago. For the past fourteen years, Professor Wolpoff has been working on the problem of human origins with colleagues in the United States, Australia, and China, uncovering fossil evidence which challenges the existence of Eve. His work and theories have been covered in *The New York Times, New Scientist, Discover,* and *Newsweek,* among other publications. He is at work on a book called *The Death of Eve.*

What About Incest?

Patrick Bateson

ir Thomas Beecham, who was a great British conductor and a man of firm opinions, is supposed to have given the following advice: "Keep clear of incest and Morris dancing." He probably regarded them both as joyless aspects of peasant life. To this day, in self-conscious attempts to re-create medieval England, men are to be seen jigging up and down and waving inflated pigs' bladders on English village greens in summer. In secret, others sexually abuse their daughters or their sisters without regard for the time of year or, indeed, rural setting.

Sexual abuse within families, about which the silence was deafening until recently, appears to be distressingly common. It may explain why prescriptions against incest are extremely widespread, existing in virtually every culture. According to this view, incest taboos protect women against men. However, if that is so, why are women not similarly protected against other forms of male violence? Dissatisfaction with a purely sociological explanation leads, commonly, to the wheeling out of a quite different argument. Incest taboos, it is suggested, are safety laws, protecting people against the biological costs of their actions. This explanation also has its problems, since most people are ill informed about the biological consequences of inbreeding. In modern societies, the biological costs are usually exaggerated grossly, but members of some societies don't seem to know about any ill effects, even though they have prohibitions against sexual relations with close kin. Even if their ancestors knew more than they do, a purely biological explanation does not account satisfactorily for many of the details of the incest taboos.

If you have access to the Anglican *Book of Common Prayer,* you will find one expression of the taboo. At the back is a table, Kindred and Affinity, "wherein whosoever are related are forbidden by the Church of England to marry together." A man may not marry his mother, sister, or daughter and a variety of other genetically related individuals. Everything follows logically for a woman and, among others, she can't marry an uncle, nephew, grandparent, or grandchild. So you might say: "Three cheers for biology!" But the list continues with the following exclusions: A man may not marry his wife's father's mother or his daughter's son's wife. The mind boggles. What was life really like in the Middle Ages if the Church felt it necessary to include these prohibitions? And did they really live so long that it was even possible to think about such marriages? But let us not get distracted. The point is that at least six of the twenty-five types of relationship that preclude marriage involve no genetic link at all— and a lot more are ambiguous on this point.

Interestingly, the Church of England does not worry about marriages between first cousins. Other cultures do, but here again, striking inconsistencies are found. In a great many cultures, marriages between parallel first cousins are forbidden, whereas marriages between cross cousins are not only allowed but, in many cases, actively encouraged. Parallel cousins are those in which the two fathers are brothers or the two mothers are sisters. Cross cousins are those in which the father of one is the brother of the other's mother. The issue, in relation to the biological explanation of incest taboos, is that parallel and cross cousins, treated so differently by the rules of many cultures, do not differ genetically.

Many animals go to considerable trouble to avoid mating with very close kin. Can we gain deeper insight into the origins of human incest taboos by looking at what is know about the evolution of inbreeding avoidance in animals? Note that "incest," which refers to a verbally transmitted prohibition in humans, should not be used for animals, if we do not wish to beg the question from the outset and assume that the human prohibition is the same as a mating inhibition in animals. Anyhow, in some animals, inbreeding can carry a heavy cost when the species has had a long history of outbreeding. For instance, if the male of a highly mobile animal like a bird is mated in the laboratory with its sister, their offspring of opposite sex are mated together and so on for several generations, the line usually dies out quickly. This is because genes usually come in pairs that may or may not be the same as their partner. Some genes are harmless when paired with a dissimilar gene but lethal when present with an identical partner. Such genes, known as lethal recessives, are more likely to be paired with an identical one as a line becomes more inbred.

Digressing briefly, these genetic costs of inbreeding, which are the ones that people usually worry about, do not provide the reason why outbreeding evolved in the first place. The presence of lethal recessive genes is a consequence of outbreeding, accumulating unseen in the genome because they are normally suppressed by a dominant partner gene. The need to outbreed in the first place is probably related to the evolution of sex. Sex is about minimizing infections and, in these terms, is worthless without outbreeding. If an animal inbreeds, it might as well multiply itself without the effort and trouble of courtship and mating. Sexual reproduction enables organisms to detect parasites much more readily than would otherwise be the case, since a parasite, disguising itself so as to be undetected by the immune systems of the host parents, is readily spotted among the tissues of the hosts' genetically distinct offspring. Without sexual reproduction in the hosts, genetic mutations on the host would also make the parasite visible to its immune system. However, parasites can quickly mutate themselves and so track changes in the host. Since they characteristically reproduce more rapidly than their hosts, escape from parasitism is difficult in species without sexual reproduction.

Now, members of many species prefer mates that are not very close kin, such as a sibling. Or, more accurately, they prefer mates that may be first or second cousins. How do they know what their cousins look like? One mechanism, which has been well studied in birds, works like this. When the birds are young, they learn about their parents and siblings, thereby providing themselves with prototypes of kin. When they are adult, they prefer slight deviations from these learned standards, thereby minimizing the ill effects both of inbreeding and of breeding with individuals that are genetically too different from themselves. It is possible to do experiments in which individual Japanese quail are brought up with non-kin. When these birds grow up, they prefer individuals that are a bit different in appearance from the ones they knew when they were young.

Natural experiments rather similar to the laboratory studies with quail have been performed on human beings. The best-analyzed case was in Taiwan in the nineteenth century and the first part of the twentieth century when it was controlled by Japan. The Japanese kept detailed records on the births, marriages, and deaths of everybody on the island. As in many other parts of Southeast Asia, the marriages were arranged and occurred mainly and most interestingly in two forms. The major type was conventional and one in which the partners met each other when adolescent. In the minor type, the wife-to-be was adopted as a young girl into the family of her future husband. In the minor type, therefore, the partners grew up together like siblings. In this way, they were treated like the quail in the laboratory experiment, having been reared with a member of the opposite sex who was not genetically related. Later in life their sexual interest in this partner was put to the test.

Arthur Wolf, an anthropologist at Stanford University, has analyzed these Japanese records meticulously. Whether he measured divorce, marital fidelity, or the number of children produced, the minor marriages were conspicuously less "successful" than the major marriages. The young couple growing up together from an early age, like brother and sister, were not especially interested in each other sexually when the time came for their marriage to be consummated. Since many explanations have been offered for this phenomenon, Wolf used the data elegantly to discriminate between the hypotheses, and he demonstrated conclusively that a major influence on human choice of sexual partners is the experience that starts in early life.

How can these aspects of human behavior help us understand the incest taboo and the different forms of the taboo
that are found around the world? To explain it, I shall introduce a speculation that was first proposed by Edward
Westermarck a century ago in his book *The History of Human
Marriage.* Westermarck suggested that if people see others
doing things that they do not do themselves, they try to stop
them. Left-handers have been feared in the past and were
forced to adopt the habits of right-handers, not simply "for
their own good." Again, gays have been given a hard time
in most cultures. In the same way, he suggested, those who
committed incest were discriminated against. People who
are familiar with members of the opposite sex from early life
are not much attracted to those individuals, and when they
spot others who are, they disapprove. It was nothing to do
with society not wanting to look after the half-witted children of inbreeding, since in many cases, they had no idea
that inbreeding had biological costs. It was all about the
suppression of disharmonizing abnormal behavior. Such conformism looks harsh to modern eyes, even though we have
plenty of examples of it in our own societies. However, when
so much depended on unity of action in the environment in
which humans evolved, wayward behavior had potentially
destructive consequences for everybody. It is not difficult to
see why conformism should have become a powerful trait in
human social behavior.

The final step in explaining the variation in incest taboos
is to suggest that people who are very familiar from early
life are the ones who are most likely to be included in the
taboo. That would account for all the strange exclusions from
the possibility of marriage in the Church of England's rules.
The explanation works particularly well for the societies in
which the parallel cousins are prohibited, but the cross cous-

ins are favored in marriage. In these cultures, the parallel cousins tend to grow up together because brothers tend to stay with brothers and sisters with sisters. The cross cousins are much less familiar with each other because brothers and sisters tend to live in different places after marriage.

If these ideas are correct, a link between biology and human incest taboos may be found, but it is not based on the simple idea that the taboo evolved, because it reinforces inbreeding avoidance and minimizes the biological costs. In essence, the proposal is that in the course of biological evolution, two perfectly good mechanisms appeared. One was concerned with getting the right balance between inbreeding and outbreeding when choosing a long-term sexual partner. The other was concerned with social conformity. When these were put together, the result was social disapproval of those who chose partners from their close family. When social disapproval was put together with an evolved language, verbal rules appeared which could be transmitted from generation to generation first by word of mouth and finally in written form. Westermarck's proposal is attractive because it helps us to understand not only why the taboos resemble each other, but also why they differ.

Sir Thomas Beecham had conventional views about incest. He was also conventional as a musician in the sense that he performed from fixed scores. Anybody familiar with a work he was conducting would have known how it would end. I have suggested that the appearance in human cultures of incest taboos was more like those forms of jazz in which the musicians improvise and develop their musical ideas from what the others have just done. As new themes emerge, the performance acquires a life of its own and ends up in a place that could not easily have been anticipated at the outset. Both the inhibition about taking a sexual partner known from early in life and social conformity were likely to be the products of biological evolution. Put them together, and humans found themselves disapproving of sex between close kin. Put that together with language, and humans had incest taboos. Few complicated processes run on railway lines.

PATRICK BATESON is professor of ethology (the biological study of behavior) at the University of Cambridge. He is also head of King's College, Cambridge, and a fellow of the Royal Society. He was trained as a zoologist and has had a long-standing research interest in the development of behavior. His interest in the human incest taboo grew out of his own work on the effects of early experience on mate choice in birds. He is author of *Measuring Behaviour* (with Paul Martin) and editor of *The Development and Integration of Behaviour.*

Why Are Some People Black?

Steve Jones

~~~~~~~~~~~~ E veryone knows—do they not?—
that many people have black skin.
What is more, black people are concentrated in certain
places—most notably, in Africa—and, until the upheavals
of the past few centuries, they were rare in Europe, Asia, and
the Americas. Why should this be so?

It seems a simple question. Surely, if we cannot give a
simple answer, there is something wrong with our understanding of ourselves. In fact, there is no straightforward
explanation of this striking fact about humankind. Its absence says a lot about the strengths and weaknesses of the
theory of evolution and of what science can and cannot say
about the past.

Any anatomy book gives one explanation of why people
look different. Doctors love pompous words, particularly if
they refer to other doctors who lived long ago. Black people
have black skin, their textbooks say, because they have a
distinctive Malpighian layer. This is a section of the skin

named after the seventeenth-century Italian anatomist Malpighii. It contains lots of cells called *melanocytes.* Within them is a dark pigment called melanin. The more there is, the blacker the skin. Malpighii found that African skin had more melanin than did that of Europeans. The question was, it seemed, solved.

This is an example of what I sometimes think of as "the Piccadilly explanation." One of the main roads in London is called Piccadilly—an oddly un-English word. I have an amusing book that explains how London's streets got their names. What it says about Piccadilly sums up the weakness of explanations that depend, like the anatomists', only on describing a problem in more detail. The street is named, it says, after the tailors who once lived there and made high collars called *piccadills.* Well, fair enough; but surely that leaves the interesting question unanswered. Why call a collar a piccadill in the first place? It is not an obvious word for an everyday piece of clothing. My book is, alas, silent.

Malphighii's explanation may be good enough for doctors, but will not satisfy any thinking person. It answers the question *how* but not the more interesting question *why* there is more melanin in African skin.

Because the parents, grandparents, and—presumably—distant ancestors of black people are black, and those of white people white, the solution must lie in the past. And that is a difficulty for the scientific method. It is impossible to check directly just what was going on when the first blacks or the first whites appeared on earth. Instead, we must depend on indirect evidence.

There is one theory that is, if nothing else, simple and consistent. It has been arrived at again and again. It depends solely on belief; and if there is belief, the question of proof does not arise. Because of this, the theory lies outside science.

It is that each group was separately created by divine action. The Judeo-Christian version has it that Adam and Eve were created in the Garden of Eden. Later, there was a gigantic flood; only one couple, the Noahs, survived. They had children: Ham, Shem, and Japhet. Each gave rise to a distinct branch of the human race, Shem to the Semites, for example. The children of Ham had dark skins. From them sprang the peoples of Africa. That, to many people, is enough to answer the question posed in this essay.

The Noah story is just a bald statement about history. Some creation myths are closer to science. They try to *explain* why people look different. One African version is that God formed men from clay, breathing life into his creation after it had been baked. Only the Africans were fully cooked— they were black. Europeans were not quite finished and were an unsatisfactory muddy pink.

The trouble with such ideas is that they cannot be disproved. I get lots of letters from people who believe passionately that life, in all its diversity, appeared on earth just a few thousand years ago as a direct result of God's intervention. There is no testimony that can persuade them otherwise. Prove that there were dinosaurs millions of years before humans, and they come up with rock "footprints" showing, they say, that men and dinosaurs lived together as friends. So convinced are they of the truth that they insist that their views appear in school textbooks.

If all evidence, whatever it is, can only be interpreted as supporting one theory, then there is no point in arguing. In fact, if belief in the theory is strong enough, there is no point in looking for evidence in the first place. Certainty is what blocked science for centuries. Scientists are, if nothing else, uncertain. Their ideas must constantly be tested against new knowledge. If they fail the test, they are rejected.

No biologist now believes that humans were created through some miraculous act. All are convinced that they evolved from earlier forms of life. Although the proof of the fact of evolution is overwhelming, there is plenty of room for controversy about how it happened. Nowhere is this clearer than in the debate about skin color.

Modern evolutionary biology began with the nineteenth-century English biologist Charles Darwin. He formed his ideas after studying geology. In his day, many people assumed that grand features such as mountain ranges or deep valleys could arise only through sudden catastrophes such as earthquakes or volcanic eruptions, which were unlikely to be seen by scientists as they were so rare. Darwin realized that, given enough time, even a small stream can, by gradually wearing away the rocks, carve a deep canyon. The present, he said, is the key to the past. By looking at what is going on in a landscape today, it is possible to infer the events of millions of years ago. In the same way, the study of living creatures can show what happened in evolution.

In *The Origin of Species,* published in 1859, Darwin suggested a mechanism whereby new forms of life could evolve. *Descent with modification,* as he called it, is a simple piece of machinery, with two main parts.

One produces inherited diversity. This process is now known as mutation. In each generation, there is a small but noticeable chance of a mistake in copying genes as sperm or egg are made. Sometimes we can see the results of mutations in skin color; one person in several thousand is an albino, lacking all skin pigment. Albinos are found all over the world, including Africa. They descend from sperm or eggs that have suffered damage in the pigment genes.

The second piece of the machine is a filter. It separates mutations which are good at coping with what the environ-

ment throws at them from those which are not. Most mutations—albinism, for example—are harmful. The people
who carry mutant genes are less likely to survive and to have
children than do those who do not. Such mutations quickly
disappear. Sometimes, though, one turns up which is better
at handling life's hardships than what went before. Perhaps
the environment is changing, or perhaps the altered gene
simply does its job better. Those who inherit it are more
likely to survive; they have more children, and the gene becomes more common. By this simple mechanism, the population has evolved through *natural selection.* Evolution,
thought Darwin, was a series of successful mistakes.

If Darwin's machine worked for long enough, then new
forms of life—new species—would appear. Given enough
time, all life's diversity could emerge from simple ancestors.
There was no need to conjure up ancient and unique events
(such as a single incident of creation) which could neither
be studied nor duplicated. Instead, the living world was itself evidence for the workings of evolution.

What does Darwin's machine tell us about skin color? As
so often in biology, what we have is a series of intriguing
clues, rather than a complete explanation.

There are several kinds of evidence about how things
evolve. The best is from fossils: the preserved remnants of
ancient times. These contain within themselves a statement
of their age. The chemical composition of bones (or of the
rocks into which they are transformed) shifts with time. The
molecules decay at a known rate, and certain radioactive substances change from one form into another. This gives a clue
as to when the original owner of the bones died. It may be
possible to trace the history of a family of extinct creatures
in the changes that occur as new fossils succeed old.

The human fossil record is not good—much worse, for

example, than that of horses. In spite of some enormous gaps, enough survives to make it clear that creatures looking not too different from ourselves first appeared around a hundred and fifty thousand years ago. Long before that, there were apelike animals which looked noticeably human but would not be accepted as belonging to our own species if they were alive today. No one has traced an uninterrupted connection between these extinct animals and ourselves. Nevertheless, the evidence for ancient creatures that changed into modern humans is overwhelming.

As there are no fossilized human skins, fossils say nothing directly about skin color. They do show that the first modern humans appeared in Africa. Modern Africans are black. Perhaps, then, black skin evolved before white. Those parts of the world in which people have light skins—northern Europe, for example—were not populated until about a hundred thousand years ago, so that white skin evolved quite quickly.

Darwin suggested another way of inferring what happened in the past: to compare creatures living today. If two species share a similar anatomy, they probably split from their common ancestor more recently than did another which has a different body plan. Sometimes it is possible to guess at the structure of an extinct creature by looking at its living descendants.

This approach can be used not just for bones but for molecules such as DNA. Many biologists believe that DNA evolves at a regular rate: that in each generation, a small but predictable proportion of its subunits changes from one form into another. If this is true (and often it is), then counting the changes between two species reveals how closely they are related. What is more, if they share an ancestor that has been dated using fossils, it allows DNA to be used as a "molecular clock," timing the speed of evolution. The rate at which the clock ticks can then be used to work out when other species split by comparing their DNA, even if no fossils are available.

Chimpanzees and gorillas seem, from their body plan, to be our relatives. Their genes suggest the same thing. In fact, each shares 98 percent of its DNA with ourselves, showing just how recently we separated. The clock suggests that the split was about six million years ago. Both chimp and gorilla have black skins. This, too, suggests that the first humans were black and that white skin evolved later.

However, it does not explain *why* white skin evolved. The only hint from fossils and chimps is that the change took place when humans moved away from the tropics. We are, without doubt, basically tropical animals. It is much harder for men and women to deal with cold than with heat. Perhaps climate has something to do with skin color.

To check this idea, we must, like Darwin, look at living creatures. Why should black skin be favored in hot and sunny places and white where it is cool and cloudy? It is easy to come up with theories, some of which sound pretty convincing. However, it is much harder to test them.

The most obvious idea is wrong. It is that black skin protects against heat. Anyone who sits on a black iron bench on a hot sunny day soon discovers that black objects heat up *more* than white ones do when exposed to the sun. This is because they absorb more solar energy. The sun rules the lives of many creatures. Lizards shuttle back and forth between sun and shade. In the California desert, if they stray more than six feet from shelter on a hot day, they die of heat stroke before they can get back. African savannahs are dead places at noon, when most animals are hiding in the shade because they cannot cope with the sun. In many creatures, populations from hot places are lighter—not darker—in color to reduce the absorption of solar energy.

People, too, find it hard to tolerate full sunshine—blacks more so than whites. Black skin does not protect those who bear it from the sun's heat. Instead, it makes the problem worse.

However, with a bit of ingenuity, it is possible to bend the theory slightly to make it fit. Perhaps it pays to have black skin in the chill of the African dawn, when people begin to warm up after a night's sleep. In the blaze of noon, one can always find shelter under a tree.

The sun's rays are powerful things. They damage the skin. Melanin helps to combat this. One of the first signs of injury is an unhealthy tan. The skin is laying down an emergency layer of melanin pigment. Those with fair skin are at much greater risk from skin cancer than are those with dark. The disease reaches its peak in Queensland, in Australia, where fair-skinned people expose themselves to a powerful sun by lying on the beach.

Surely, this is why black skin is common in sunny places—but, once again, a little thought shows that it probably is not. Malignant melanoma, the most dangerous skin cancer, may be a vicious disease, but it is an affliction of middle age. It kills its victims after they have passed on their skin-color genes to their children. Natural selection is much more effective if it kills early in life. If children fail the survival test, then their genes perish with their carriers. The death of an old person is irrelevant, as their genes (for skin color or anything else) have already been handed on to the next generation.

The skin is an organ in its own right, doing many surprising things. One is to synthesize vitamin D. Without this, children suffer from rickets: soft, flexible bones. We get most vitamins (essential chemicals needed in minute amounts) from food. Vitamin D is unusual. It can be made in the skin by the action of sunlight on a natural body chemical. To do this, the sun must get into the body. Black people in sunshine hence make much less vitamin D than do those with fair skins. Vitamin D is particularly important for children, which is why babies (African or European) are lighter in color than are adults.

Presumably, then, genes for relatively light skin were favored during the spread from Africa into the cloud and rain of the north. That might explain why Europeans are white—but does it reveal why Africans are black? Too much vitamin D is dangerous (as some people who take vitamin pills discover to their cost). However, even the fairest skin cannot make enough to cause harm. The role of black skin is not to protect against excess Vitamin D.

It may, though, be important in preserving other vitamins. The blood travels around the body every few minutes. On the way, it passes near the surface of the skin through fine blood vessels. There, it is exposed to the damaging effects of the sun. The rays destroy vitamins—so much so, that a keen blond sunbather is in danger of vitamin deficiency. Even worse, the penetrating sunlight damages antibodies, the defensive proteins made by the immune system. In Africa, where infections are common and, sometimes, food is short, vitamin balance and the immune system are already under strain. The burden imposed by penetrating sunlight may be enough to tip the balance between health and disease. Dark skin pigment may be essential for survival. No one has yet shown directly whether this is true.

There are plenty of other theories as to why some people are black. For an African escaping from the sun under a tree, black skin is a perfect camouflage. Sexual preference might even have something to do with the evolution of skin color. If, for one reason or another, people choose their partners on the basis of color, then the most attractive genes will be passed on more effectively. A slight (and perhaps quite accidental) preference for dark skin in Africa and light in Europe would be enough to do the job. This kind of thing certainly goes on with peacocks—in which females prefer males with brightly patterned tails—but there is no evidence that it happens in humans.

Accident might be important in another way, too. Probably only a few people escaped from Africa a hundred thousand years and more ago. If, by chance, some of them carried genes for relatively light skins, then part of the difference in appearance between Africans and their northern descendants results from a simple fluke. There is a village of North American Indians today where albinos are common. By chance, one of the small number of people who founded the community long ago carried the albino mutation and it is still abundant there.

All this apparent confusion shows how difficult it is for science to reconstruct history. Science is supposed to be about testing, and perhaps disproving, hypotheses. As we have seen, there is no shortage of ideas about why people differ in skin color. Perhaps none of the theories is correct; or perhaps one, two, or all of them are. Because whatever gave rise to the differences in skin color in different parts of the world happened long ago, no one can check directly.

But science does not always need direct experimental tests. A series of indirect clues may be almost as good. The hints that humans evolved from simpler predecessors and are related to other creatures alive today are so persuasive that it is impossible to ignore them. So far, we have too few facts and too many opinions to be certain of all the details of our own evolutionary past. However, the history of the study of evolution makes me confident that, some day, the series of hints outlined in this essay will suddenly turn into a convincing proof of just why some people are black and some white.

STEVE JONES is a biologist and professor in genetics at University College, London, and head of the Department of Genetics and Biometry at the Galton Laboratory at UCLA. His research interests are in the genetics of the evolutionary process in animals, from fruit flies to humans. Much of his fieldwork has been on the ecological genetics of snails, but he has also published many papers on human genetics and evolution. He has been particularly involved in the mathematical analysis of patterns of genetic change in modern humans in relation to fluctuations in population size, but has also published many papers on the genetic implications of the human fossil record and on the biological nature of the human race.

Professor Jones is a regular contributor to the journal *Nature*'s "News and Views" section and to the BBC radio science programs on Radio 4 and the World Service. He is author of *The Language of the Genes*; and coeditor (with Robert Martin and David Pilbeam) of *The Cambridge Encyclopedia of Human Evolution.*

Chance and the History of Life

Peter Ward

I t is a common human trait to take the familiar for granted. Take, for example, the wondrous diversity of mammals now living on our planet. The forests, grasslands, oceans, even the skies have been colonized and in a sense conquered by mammalian species. There are so many mammals, and they are so dominant, the geologists have long called the present geological era the Age of Mammals. And yet, the so-called Age of Mammals has been the reigning order of things for but a small fraction of the time that life has existed on this earth. Indeed, the fossil record tells us that mammals have been around for two hundred and fifty million years but dominant for only the last fifty million years of that great expanse of time. Perhaps we should ask: Why has the Age of Mammals lasted such a short time, rather than for so long, or why is it that mammals (including humans) now rule the earth at all?

Good mammals ourselves, we have all been long indoctrinated with the perhaps chauvinistic notion that mammals are the dominant members of the land fauna because we are superior to the other classes of land vertebrates—birds, reptiles, and amphibians. Birds are clearly the rulers of the air, but only a few exist as purely land-living forms. Reptiles and amphibians are relatively few in number compared with mammals in most terrestrial habitats. Most scientists agree that the aspects of "mammalness" that impart the great competitive superiority enjoyed by mammals are warm-blooded metabolism, a highly developed brain, a good fur overcoat to withstand cold winters, and parental care of the young.

When I was in school, the story of the rise of mammals went something like this: The first land-living creatures with backbones crawled from the sea (or lakes, streams, and ponds) about four hundred million years ago. These earliest terrestrial pioneers were amphibians. They, in turn, were the evolutionary ancestors of reptiles, who were distinguished by the ability to lay eggs and grow from juveniles to adults in a fully terrestrial habitat, as well as changes in skeletal architecture from the original amphibian design. These changes gave rise to the first true mammals somewhere about 225 million years ago. The early mammals, however, did not immediately take over the earth but had to spend a rather lengthy apprenticeship to the then-incumbent terrestrial overlords, the dinosaurs. Eventually, quality rises to the top, and the dinosaurs became obsolete; and, due to pressures of climate change and the onslaught of hordes of little egg-eating mammals, dinosaurs went extinct against a backdrop of exploding volcanoes. In this scenario the dinosaurs had the good grace to hand over the stewardship of the continents to creatures better adapted to living on land. Therefore, mammals now control the earth because they are inherently superior physiologically and ecologically. Or so the story goes.

But let us reexamine this scenario by conducting a pale-ontological thought experiment. Thought experiments have long been a favorite exercise of physicists; Albert Einstein loved them. In our paleontological thought experiment, we are going to have as much fun as any physicist ever had; we are going to reintroduce dinosaurs into California.

The beauty of thought experiments is that you don't have to worry about your budget, so let us spare no expenses. Let us build a giant fence around Mendocino County, California, and introduce a seventy-million-year-old, or Late Cretaceous, dinosaur fauna into our new park. (Shall we call it Cretaceous Park?) In our park, we now have big dinosaurs, such as tri-ceratops, lots of duck-billed dinosaurs, a few long-necked sauropods (since they were largely gone by the Late Creta-ceous, we only have a few), lots of little dinosaurs with un-pronounceable names I always forget, and, of course, a few large carnivorous forms such as the inevitable *T. rex.* Because Mendocino County has some Late Cretaceous forests still in-tact (redwood trees are living fossils dating back to the time of dinosaurs), there would be plenty of familiar food for our herbivorous dinosaurs; it is not only possible but probable that our new California immigrants would not only survive but thrive.

But our park is not solely populated by dinosaurs; we are leaving the native mammalian populations in there as well, such as deer, elk, porcupines, bobcats, bears, and a mountain lion or two. Now, we go away for ten thousand years and leave this mix from the Age of Reptiles and the Age of Mammals to do their thing. What species would we find when we came back? Would mammals, with their traits of parental care and mammalian warm-bloodedness win out, driving the dinosaurs to extinction through competition for space and food? Or would the dinosaurs hold their own?

Could there be some curious but wonderful combination of mammals and dinosaurs left at the end? My guess is that the dinosaurs would not only successfully compete with mammals for food and space, but they might just wipe out most of the larger mammals over the course of time. Perhaps our Mendocino dinosaurs would even escape from their California prison, and like good Californians of today, move to more "livable" places, like Seattle or Pittsburgh, eventually ruling the earth once more. Michael Crichton may be right: We had better keep the damn things off the mainland at any cost.

Now let us repeat our imaginary experiment a hundred times. Would the assemblage of animals still alive at the end of the experiment be the same each time? Would the experiment repeat itself in the same fashion at each iteration? I think it most unlikely. Like human history, the history of life is apparently susceptible to minor actions or influences, which amplify into gigantic changes over the passage of time.

So why are dinosaurs no longer around, if they were so well adapted? The answer is that chance, not determinism, has orchestrated the development of evolution. The death of the dinosaurs may be the best of all examples. Sixty-five million years ago, the earth was struck by an asteroid of enormous size, which left a crater almost two hundred miles in diameter, in the Yucatan region of Mexico. The odds against being hit by such a large celestial body are staggering, but hit we were. The dinosaurs had been carrying on quite nicely for over a hundred million years prior to this asteroid strike. This giant catastrophe tidily removed all of the dinosaurs, and paved the way for the evolution of today's Age of Mammals. Had that great asteroid impact not taken place, we would surely not have the fauna of animals and plants on earth today, and there is a very good chance that the dominant land animals would still be dinosaurs. And if that were the case, would our species have ever evolved? I think not.

The role of chance in the history of life may have affected the evolution of human culture as well. During the summer of 1993, an astounding discovery was made in Greenland. Analyses of ice cores showed that the last ten thousand years on earth has been a time of relatively little climate change. Prior to this long period of climate stability, the earth had been subject to short and sudden swings of climate, leading to rapid glaciations followed by warmer intervals. Global temperature changes of as much as ten to fifteen degrees may have taken place over time intervals of decades, rather than millennia, as formerly believed. Some scientists (myself included) now suspect that the rise of agriculture and human civilization is very much a result of the period of climate stability we are now experiencing. Our species has been on earth for over one hundred thousand years, but only in the last five thousand have we mastered agriculture and built cities. What were we up to for the other ninety-five thousand years? Were we like the early mammals, trembling not before behemoth dinosaurs but slave to enormous swings in climate? Once again, chance, in the form of a rare period of climate stability, has affected the history of life on this planet.

Our species, now so numerous and growing more so each day, will surely affect the biota of our planet as severely as the great asteroid that wiped out the dinosaurs. We humans are ourselves a chance event that will surely put an end to the Age of Mammals as we now know it. The irony of the whole situation is that God's latest throw of the dice, at least for the Age of Mammals, may have come up snake eyes.

PETER DOUGLAS WARD is professor of geological sciences, professor of zoology, curator of paleontology, and chairman of geology and paleontology at the Burke Museum, University of Washington in Seattle. He also holds an adjunct faculty position at the California Institute of Technology at Pasadena.

He has had a rather schizophrenic scientific career, dividing his time between studies on the fossil record and research into the nature of modern marine communities. Since earning his Ph.D. in 1976, Ward has published more than eighty scientific papers dealing with these topics. He currently chairs an international panel on the Cretaceous-tertiary extinction, and served as editor of the recently published volume *Global Catastrophes in Earth History,* sponsored by the National Academy of Science and NASA. He is an advisor to the Lunar and Planetary Institute (NASA) on extinction in the geological record. He was elected as a Fellow of the California Academy of Science in 1984, and has been nominated for the Schuchert Medal, awarded by the Paleontological Society. He was recently elected as senior councilor to the Paleontological Society, and is involved in creating a new natural history museum in Seattle, Washington.

Ward has written five books including *In Search of Nautilus; On Methuselah's Trail;* and *The End of Evolution*—the latter two were nominees for the Los Angeles Times Book Prize (science category), and *On Methuselah's Trail* also received the Golden Trilobite Award from the Paleontological Society as best popular science book for 1992.

Nobody Loves a Mutant

Anne Fausto-Sterling

The Reverend John Mather (writing in 1636) and Spider-Man (speaking in 1994) would disagree. Mather referred to sodomy as "unnatural filthiness," while Spider-Man exclaims that "nature *proves* God never intended us to be all the same." Each turns to biology to justify a belief about how humans ought to be. Mather judged an unnatural behavior severely, suggesting sodomy be named a capital crime. A bit more sympathetic to the plight of the comic-strip mutant, hated and feared by just normal folk, Spider-Man uses nature to account for the normality of human difference. Implicitly and explicitly both make use of and confound two basic concepts in biology and medicine—the normal and the natural.

From the vantage point of the evolutionary biologist, reproduction is the stuff of evolution. What, then, could be less evolutionarily valuable, more unnatural, than a sexual coupling that precludes offspring? How can it be that in the latter half of the twentieth century, behavioral biologists have documented many cases of same-sex coupling in animals? Noted behavioral biologist Frank Beach detailed the commonness of both male-male and female-female mounting. In mammals, such behaviors occur in at least thirteen different species coming from as many as five different taxonomical orders. Sometimes same-sex interactions in the animal world serves a nonreproductive social function. If we expanded our definition of normality beyond the question of reproduction, we could easily call such interactions natural.

Moreover, sometimes same-sex coupling in animals *does* relate to reproduction. Living in the American Southwest is an all-female race of lizards. They reproduce parthenogenetically—the female produces an egg which manages to maintain or restore its proper chromosome number and to kick start embryonic development without the help of a sperm. Biologist David Crews noticed with more than a little surprise that these females copulated. One female changed color to look more like the males found in heterosexual populations, and mounted the other. Instead of inserting a penis, however, they rubbed cloacae. The mounted female would then lay eggs—many more than if she had not copulated. After some time elapsed, the mounting female lost her male-like characteristics, her ovary developed ready-to-ovulate eggs, and she might then be mounted by a "male" female who had already laid all of her mature eggs. Crews has shown in some detail how this homosexual behavior enhances reproductive success by increasing egg-laying.

Why discuss lesbian lizards? I am not arguing that human homosexuality evolved from animal homosexuality (although it might have), nor am I suggesting that same-sex contacts among humans are mere inborn animal urges. My point is far simpler: Since homosexuality occurs in the biological world under a variety of circumstances, one cannot use nature to argue that it is unnatural. If nature provides the gold standard of what is natural, homosexuality and heterosexuality are equally golden.

We can score one for Spider-Man, but the battle doesn't end there. Were John Mather with us today, he might rise to his own defense. "Unnatural," he might opine "was just a seventeenth-century way to say abnormal." What he really meant to say was that whether or not you can find animal examples, homosexuality is abnormal behavior. Now we have leaped away from an argument based on simple example— either animals do it or they don't—and have launched ourselves into two different sorts of claims, one moral and the other statistical.

The moral claim inevitably must appeal to God or some other form of higher order. Mather might have argued that God never meant humans to engage in anything other than reproductive sex, and he would have defined intercourse for the purpose of creating children as the moral norm. Other behaviors would be abnormal in the eyes of God. With this turn of the argument, nature and the natural become irrelevant. Spider-Man can only retort with a counterclaim about what God meant nature to show us. Once the debate is placed on these moral grounds, of course, we have abandoned scientific argument in favor of theological ones. And theological arguments run into big trouble if they seek justification in the natural world. Rather, they must be based on belief which can only be substantiated by faith. On these terms, lesbian lizards are patently irrelevant.

While biologists and medical scientists don't usually call a phenomenon unnatural, they are happy to refer to it as abnormal. In doing so, they move from a presence/absence argument to a statistical one. Consider the inherited trait called polydactyly. It occurs rarely in the human population and causes the birth of a child with more than the usual ten fingers and ten toes. Physicians consider it to be abnormal— the normal child having twenty digits. Often a doctor will surgically remove an extra finger or two so that the child will look and feel "normal" (there's that problematic word again).

Polydactyly is natural because it is known in nature, but it is statistically unusual. If you took a sample of one million children and counted their digits at birth, you would find that the vast majority of them had twenty in all, while some might have more and some less. If you displayed this digital variation graphically, you would come up with what statisticians call a normal curve. On paper, it looks like a drawing of the Liberty Bell (minus the crack), and so is called a bell curve. Most of the area underneath the curve is devoted to people with 19, 20, or 21 digits. But smaller areas—the so-called tails—would represent the small number of kids with 16, 17, 18 (or fewer) and 22, 23, or 24 (or more) digits. A practicing doctor would call all of the ten-toe-ten-finger kids normal. In fact, they are a statistical mean. A doc would label the others abnormal and refer to their birth defects.

The matter may sound clear-cut when one is talking about fingers and toes. But what about more variable traits, such as height? How can we decide when someone is so small we call him a midget, or so tall we think she is a giant? Common medical practice relies on a procedure that is arbitrary at its heart—calculating the standard deviation, a measure of the amount of variation around the mean or full height of the bell curve. Medical statisticians often label as abnormal anything that is greater or less than two times the standard deviation. Clearly, such cases are way out there at the edges of the normal curve—the two foot and the seven-and-a-half-foot adult might count as midgets or giants—abnormal in their final adult heights.

From a practical point of view, such statistical abnormality might matter quite a lot. Designers, after all, build beds, water fountains, cars, kitchen sinks, toilets, and ceilings to cater to the mean and its close neighbors. Somebody who is seven and a half feet tall needs specially built beds and personally tailored clothing. A person born with eight toes on each foot may also have a hard time walking, and surgeons might operate to make the foot more normal. What starts out as a statistical difference becomes a social or medical problem. The infrequent becomes the abnormal.

Humans have learned how to change nature. Recently, for example, medical researchers discovered how to turn bacteria into factories that can make human growth hormone (responsible for long bone growth and, thus, height in children) in large quantities. In the old days, when growth hormone had to be extracted from human pituitary glands (which were, of course, in short supply), only children whose potential height fell on the very short side of the tail of the normal curve received treatment. This helped some who would have remained as midgets to reach a closer-to-the-average height. But there's no money in marketing growth hormone to potential midgets. There just aren't enough of them to go around. So doctors (and the National Institutes of Health) have defined a new disease called "short stature." It is a statistical disease; it also seems to be of greater concern to males than females. For this new disease, boys who are unlikely to grow to more than five feet become eligible for growth hormone treatment (to make them normal, of course). But the new treatment will also change the bell curve, the normal distribution that defines short stature in the first place.

By changing the measures that fall under the standard curve, are we creating something unnatural? If the use of growth hormone becomes sufficiently common, we will make a new population with altered height distributions, one which previously did not exist in nature. And is it right to do this? Here, once more, we revert to the moral question: Are there some things that God or the Great Spirit, or however one expresses one's spiritual sense, just didn't intend us to do? Are short adults abnormal? Is it morally wrong to deny them the medicine that will "cure" their disease? Are homosexuals unnatural? If not, are *they* abnormal? The unnatural, the natural, the normal, the abnormal, the moral, and immoral slip back and forth. When we discuss them, let's at least keep our footing as precise as possible. Then, at least, we can all argue about the same thing at the same time. John Mather and Spider-Man may never come to an understanding. But they can be clearer about where their paths diverge.

ANNE FAUSTO-STERLING is professor of medical science in the division of biology and medicine at Brown University, Providence, Rhode Island. She is a fellow of the American Association for the Advancement of Science and has been the recipient of grants and fellowships in both the sciences and the humanities.

In addition, as author of scientific publications in the field of *Drosophila* (fruit fly) developmental genetics, she has written several papers that critically analyze the role of preconceptions about gender in structuring theories of development.

Professor Fausto-Sterling has also written more broadly about the role of race and gender in the construction of scientific theory and the role of such theories in the construction of ideas about race and gender. She is author of *Myths of Gender: Biological Theories about Women and Men,* and is currently working on two books.

The first deals with biology and the social construction of sexuality, while the second examines the intersections of race and gender in the founding moments of modern biology. She is a strong advocate of the idea that understanding science is of central importance to feminist students and scholars, and that understanding feminist insights into science is essential to science students and researchers.

Part Four

Mind

How to Make Mistakes

Daniel C. Dennett

〰〰〰〰〰〰 **M**aking mistakes is the key to making progress. There are times, of course, when it is important not to make any mistakes—ask any surgeon or airline pilot. But it is less widely appreciated that there are also times when making mistakes is the secret of success. What I have in mind is not just the familiar wisdom of nothing ventured, nothing gained. While that maxim encourages a healthy attitude toward risk, it doesn't point to the positive benefits of not just risking mistakes, but actually of making them. Instead of shunning mistakes, I claim, you should cultivate the habit of making them. Instead of turning away in denial when you make a mistake, you should become a connoisseur of your own mistakes, turning them over in your mind as if they were works of art, which in a way, they are. You should seek out opportunities to make grand mistakes, just so you can then recover from them.

First, the theory, and then, the practice. Mistakes are not

just golden opportunities for learning; they are, in an important sense, the only opportunity for learning something truly new. Before there can be learning, there must be learners. These learners must either have evolved themselves or have been designed and built by learners who evolved. Biological evolution proceeds by a grand, inexorable process of trial and error—and without the errors, the trials wouldn't accomplish anything. This is true wherever there is a design process, no matter how clever or stupid the designer. Whatever the question or design problem is, if you don't already know the answer (because someone else figured it out already and you peeked, or because God told you), the only way to come up with the answer is to take some creative leaps in the dark and be informed by the results. You, who know a lot—but just not the answer to the question at hand—can take leaps somewhat guided from the outset by what you already know; you may not be just guessing at random.

For evolution, which knows nothing, the leaps into novelty are blindly taken by mutations, which are copying "errors" in the DNA. Most of these are fatal errors, in fact. Since the vast majority of mutations are harmful, the process of natural selection actually works to keep the mutation rate very low. Fortunately for us, it does not achieve perfect success, for if it did, evolution would eventually grind to a halt, its sources of novelty dried up. That tiny blemish, that "imperfection" in the process, is the source of all the wonderful design and complexity in the living world.

The fundamental reaction to any mistake ought to be this: "Well, I won't do *that* again!" Natural selection takes care of this "thought" by just wiping out the goofers before they can reproduce. Something with a similar selective force— the behaviorists called it "negative reinforcement"—must operate in the brain of any animal that can learn not to make that noise, touch that wire, or eat that food. We human beings carry matters to a much more swift and efficient level. We can actually think the thought, reflecting on what we have just done. And when we reflect, we confront directly the problem that must be solved by any mistake maker: What, exactly, is *that?* What was it about what I just did that got me into all this trouble? The trick is to take advantage of the particular details of the mess you've made, so that your next attempt will be informed by it, and not be just another blind stab in the dark. In which direction should the next attempt be launched, given that this attempt failed?

At its simplest, this is a technique we learned in grade school. Recall how strange and forbidding long division seemed at first: You were confronted by two imponderably large numbers, and you had to figure out how to start. Does the divisor go into the dividend six or seven or eight times? Who knew? You didn't have to know; you just had to take a stab at it, whichever number you liked, and check the result. I remember being almost shocked when I was told I should start by just "making a guess." Wasn't this *mathematics?* You weren't supposed to play guessing games in such a serious business, were you? But eventually, I came to appreciate the beauty of the tactic. If the chosen number turned out to be too small, you increased it and started over; if too large, you decreased it. The good thing about long division was that it always worked, even if you were maximally stupid in making your first choice, in which case, it just took a little longer.

This general technique of making a more-or-less educated guess, working out its implications, and using the result to make a correction for the next phase has found many applications. Navigators, for instance, determine their position at sea by first making a guess about where they are. They make a guess about *exactly*—to the nearest mile—what their latitude and longitude are, and then work out how high in the sky the sun would appear to be if that were (by an incredible coincidence) their actual position. Then they measure the actual elevation angle of the sun, and compare the two values. With a little more trivial calculation, this tells them how big a correction, and in what direction, to make to their initial guess. It is useful to make a good guess the first time, but it doesn't matter that it is bound to be mistaken; the important thing is to make the mistake, in glorious detail, so you have something serious to correct.

The more complex the problem, of course, the more difficult the analysis is. This is known to researchers in artificial intelligence (AI) as the problem of "credit assignment" (it could as well be called blame assignment). Many AI programs are designed to "learn," to adjust themselves when they detect that their performance has gone awry, but figuring out which features of the program to credit and which to blame is one of the knottiest problems in AI. It is also a major problem—or at least a source of doubt and confusion—in evolutionary theory. Every organism on earth dies sooner or later after one complicated life story or another. How on earth could natural selection see through the fog of all these details in order to discern the huge complex of positive and negative factors and "reward" the good and "punish" the bad? Can it really be that some of our ancestors' siblings died childless because their eyelids were the wrong shape? If not, how could the process of natural selection explain why our eyelids came to have the nifty shape that they do?

One technique for easing the credit assignment problem is to build mistake opportunities into a "hierarchy"—a sort of pyramid of levels, with a safety net at each step. By and large, don't mess with the parts that are already working well, and take your risks opportunistically. That is, plan your project so that at each step you can check for error and take a remedial path. Then you can be bold in execution, ready to take advantage of unlikely success and ready to cope gracefully with likely failure. This is a technique that stage magicians—at least the best of them—exploit with amazing results. (I don't expect to incur the wrath of the magicians for revealing this trick to you, since this is not a particular trick but a deep general principle.) A good card magician knows many tricks that depend on luck—they don't always work, or even often work. There are some effects—they can hardly be called tricks—that might work only once in a thousand times! But here is what you do. You start by telling the audience you are going to perform a trick, and without telling them what trick you are doing, you go for the one-in-a-thousand effect. It almost never works, of course, so you glide seamlessly into a second try, for an effect that works about one time in a hundred, perhaps. When it too fails (as it almost always will), you slide into effect number three, which only works about one time in ten, so you'd better be ready with effect number four, which works half the time (let's say), and if all else fails (and by this time, usually one of the earlier safety nets will have kept you out of this worst case), you have a failsafe effect, which won't impress the crowd very much but at least it's a surefire trick. In the course of a whole performance, you will be very unlucky indeed if you always have to rely on your final safety net, and whenever you achieve one of the higher-flying effects, the audience will be stupefied. "Impossible! How on earth could you have known that was my card?" Aha! You didn't know, but you had a cute way of taking a hopeful stab in the dark that paid off. By hiding the "error" cases from view, you create a "miracle."

Evolution works the same way: all the dumb mistakes tend to be invisible, so all we see is a stupendous string of triumphs. For instance, over 90 percent of all the creatures that have ever lived died childless, but not a single one of your ancestors suffered that fate. Talk about a line of charmed lives!

The main difference between science and stage magic is that in science, you make your mistakes in public. You show them off, so that everybody—not just yourself—can learn from them. This way, you get the benefit of everybody else's experience, and not just your own idiosyncratic path through the space of mistakes. This, by the way, is what makes us so much smarter than every other species. It is not so much that our brains are bigger or more powerful, but that we share the benefits that our individual brains have won by their individual histories of trial and error.

The secret is knowing when and how to make mistakes, so that nobody gets hurt and everybody can learn from the experience. It is amazing to me how many really smart people don't understand this. I know distinguished researchers who will go to preposterous lengths to avoid having to acknowledge that they were wrong about something—even something quite trivial. What they have never noticed, apparently, is that the earth does not swallow people up when they say, "Oops, you're right. I guess I made a mistake." You will find that people love pointing out your mistakes. If they are generous-spirited, they will appreciate you more for giving them the opportunity to help, and acknowledging it when they succeed, and if they are mean-spirited, they will enjoy showing you up. Either way, you—and we all—win.

Of course, people do not enjoy correcting the *stupid* mistakes of others. You have to have something bold and interesting to say, something original to be right or wrong about, and that requires building the sort of pyramid of risky thinking we saw in the card magician's tricks. And then there's a surprise bonus: If you are one of the big risk takers, people will even get a kick out of correcting your stupid mistakes, which show that you're not so special, you're a regular bungler like the rest of us. I know philosophers who have never—apparently—made a mistake in their work. Their specialty is pointing out the mistakes of others, and this can be a valuable service, but nobody excuses *their* errors with a friendly chuckle.

We don't usually have to risk life and limb in order to learn from our mistakes, but we do have to keep track, and actually attend to them. The key to that is first, not to try to hide your mistakes. If you hide them, you may, like the magician, enhance your reputation, but this is a short-range solution that will come to haunt you in the long run. Second, you must learn not to deny *to yourself* that you have made them or try to forget them. That is not easy. The natural human reaction to mistake is embarrassment and anger, and you have to work hard to overcome these emotional reactions. Try to acquire the weird practice of savoring your mistakes, delighting in uncovering the strange quirks that led you astray. Then, once you have sucked out all the goodness to be gained from having made them, you can cheerfully forget them, and go on to the next big opportunity.

You are going to make lots of mistakes in your life, and some of them, unless you truly do lead a charmed life, will really hurt—yourself and others. There are some ways of making the best of it, since the more you learn from the relatively painless mistakes, the less likely you are to commit the awful variety.

PHILOSOPHER DANIEL C. DENNETT is director of the Center for Cognitive Studies, and distinguished arts and sciences professor at Tufts University in Medford, Massachusetts. He is a member of the American Academy of Arts and Sciences and the Academia Scientiarum et Artum Europaea. His books have made his name well known to a wide reading public. *Brainstorms* was hailed by Douglas Hofstadter in *The New York Review of Books* as "one of the most important contributions to thinking about thinking yet written." This led Dennet to seek Hofstadter out as his collaborator on *The Mind's I,* which has sold fifty thousand copies in hardback and several hundred thousand in paperback. Most recently, Dennett is the author of *Consciousness Explained* and the forthcoming *Darwin's Dangerous Idea: Evolution and the Meanings of Life.*

Professor Dennett is in great demand as a public lecturer at colleges and universities, and is frequently called by *Time, Newsweek, The New York Times, The Wall Street Journal,* and other national periodicals for comments and advice on news stories on the mind, brain, computers, and artificial intelligence. He is chairman of the prize committee for the hundred-thousand-dollar Loebre Prize, to be awarded to the first AI program that wins the unrestricted Turing Test.

Can Minds Do More Than Brains?

Hao Wang

C an minds do more than brains? Is this a scientific question?

Scientists and philosophers today generally assume that minds and brains are equivalent, that there is a one-to-one correspondence between the mental states and physical states of one's brain. A recognized scientific term for this assumption is "psychophysical parallelism," which both Kurt Gödel and Ludwig Wittgenstein regarded as a prejudice of our time. Of course, a prejudice is not necessarily false. Rather, it is a strongly held belief which is not warranted by the available evidence, the intensity of the conviction being disproportionate to the evidence for holding it.

Locating the site of mental activities is not obvious from everyday experience. For instance, the same Chinese character *hsin* or *xin,* often translated mind-heart or body-mind, stands for both the mind and the heart, suggesting that mental activities take place in the organ of the heart. The Greek philosopher Empedocles also thought that the mind is located in the heart, taken as the material organ. Of course, nowadays we are aware of the existence of empirical evidence indicating that the brain is directly associated with one's mental activities. For example, if some parts of the brain are damaged or removed or somehow disconnected, certain mental activities cease.

But the available evidence is far from proving that every mental state corresponds to a unique physical state in the brain. For all we know, certain subtle changes in the mind may correspond to no physical changes in the brain. That is why psychophysical parallelism is, relative to our present knowledge, only an assumption. We seem able to observe more mental differences than physical differences in the brain. For instance, if we think of the spectacular mental feats of Franz Schubert or Albert Einstein, we are far from being in a position to discern enough physical differences between their brains and some physically similar brains to give an adequate account of the drastic difference in their respective outputs.

Even in ordinary experience, it is difficult to exclude the possibility that our minds make more distinctions than our brains. For example, I have tried to reconstruct my discussions with Gödel in the 1970s with the help of certain rough notes that I made at the time. These notes reminded me of many things. Presumably, my brain contains certain physically discernible traces which are not represented in actual notes. But it is also possible that my mind remembers more than is embodied in these traces in the brain.

Wittgenstein suggested a thought experiment which seems to give a more precise formulation for such a possibility. He imagined someone making jottings, as a text is being recited, that are necessary and sufficient for that person to reproduce the text. He continued: "What I called jottings would not be a rendering of the text, not so to speak a translation with another symbolism. The text would not be stored up in the jottings. And why should it be stored up in our nervous system?" In other words, certain reminders stored on paper are, as we know, often sufficient to bring about a correct reproduction. For all we know, this may also be the case with the brain: namely, that it may contain just sufficient reminders rather than counterparts of everything in the mind.

One reason for the belief in psychophysical parallelism is undoubtedly an inductive generalization from the great success of physics—not only in dealing with physical phenomena, but also in dealing with biological issues, notably on the molecular level. Indeed, as science advances, we tend to detect closer and closer correlation between mental and neural events. Yet, given our experience with the striking distinction between the physical and the mental, it is by no means clear that, of the comprehensive assertion of parallelism, how typical a part our limited knowledge of correlation makes up.

At the same time, we are struck by the contrast between the maturity of our study of the physical world and the primitive state of our attempts to deal directly with mental phenomena. There is a natural inclination to equate "science" with the science of the physical world. Consequently, one is inclined to believe that, if mind cannot be explained in terms of the science of the physical world, then it cannot be explained in any scientific terms at all. Given such a belief, psychophysical parallelism becomes a necessary condition for the possibility of a scientific treatment of mental phenomena.

Even though we have so far failed to study mental phenomena nearly as systematically as physical phenomena, this fact in itself is not proof that parallelism is true, or even irrefutable. We do not know whether or not we will be able to study mental phenomena directly in an effective manner.

In our discussions in 1972, Gödel conjectured that parallelism is false, and that it will be disproved scientifically—perhaps by the fact that there aren't enough nerve cells to perform the observable operations of the mind. In this context, observable operations undoubtedly include the operations of one's memory, reflection, imagination, etc., which are directly observable only by introspection. As we know, what is observable by introspection can often be communicated to others in such a way that they can also test the observations with the help of analogous introspective observations. There is no reason why introspective evidence should be excluded altogether.

The conjecture about there being not enough nerve cells to perform the observable mental operations illustrates significantly a test of what we mean by a problem being "scientific." Certainly, the capacity of nerve cells is a natural and central topic of study for neuroscience. At the same time, the observable operations of the mind are also things which we are capable of knowing. These facts are undoubtedly the reason why our initial response is to agree with Gödel that the conjecture is indeed a scientific one, even though many people would guess that the conjectured conclusion is false rather than true.

What is attractive about the conjecture is the apparent sharpness of its quantitative flavor. But the number of neurons in a brain is estimated at 10^{11} or 10^{12}, and there are many more synapses than neurons. We are not accustomed to dealing with the implications of such large numbers of units, and our knowledge of their actual capacities—in particular, of the expected gap between their combinationally possible and their actually realizable configurations—is very limited indeed. Moreover, as we know, the mind increases its power by tools such as pencil, paper, computers, etc., by learning from others; and by using books and one's own writings as a kind of external memory.

We have little knowledge of the extent to which the brain does the same sort of thing, too. Therefore, the situation with the brain may be less clear than Gödel seems to think. At the same time, we are at present far from possessing any sort of fully promising idea to guide us into the quantitative determination of all the observable operations of the mind.

Even though we are at present far from being able to decide whether Gödel's conjecture is true or false, it is hard to deny that it is a meaningful—and scientific—conjecture. Indeed, the conjecture seems to me to illustrate the possibility of a definitive resolution of the perennial controversy between materialism and idealism, which has been considered so important in so many different ways.

HAO WANG, logician, has been professor of logic at The Rockefeller University in New York for nearly three decades. From 1961 to 1967, he was Gordon McKay Professor of Mathematical Logic and Applied Mathematics at Harvard University, and from 1955 to 1961, John Locke Lecturer in Philosophy, then reader in the philosophy of mathematics, at the University of Oxford. He is the author of many articles and several books on logic, computers, and philosophy including *From Mathematics to Philosophy; Beyond Analytic Philosophy; Reflections on Kurt Gödel;* and *Computation, Logic, Philosophy.*

How to Think What No One Has Ever Thought Before

William H. Calvin

⁓⁓⁓⁓⁓⁓ **T**he short answer is to take a nap and dream about something. Our dreams are full of originality. Their elements are all old things, our memories of the past, but the combinations are original. Combinations make up in variety what they lack in quality, as when we dream about Socrates driving a bus in Brooklyn and talking to Joan of Arc about baseball. Our dreams get time, place, and people all mixed up.

Awake, we have a stream of consciousness, also containing a lot of mistakes. But we can quickly correct those mistakes, usually before speaking out loud. We can improve the sentence, even as we are speaking it. Indeed, most of the sentences we speak are ones we've never spoken before. We construct them on the spot. But *how* do we do it, when we say something we've never said before—and it doesn't come out as garbled as our dreams?

We also forecast the future in a way that no other animal can do. Since it hasn't happened yet, we have to imagine what might happen. We often preempt the future by taking actions to head off what will otherwise happen. We can think before acting, guessing how objects or people might react to a proposed course of action.

That is extraordinary when compared to all other animals. It even takes time to develop in children. By the time children go to school, adults start expecting them to be responsible for predicting the consequences: "You should have realized that . . ." and "Think before you do something like that!" aren't seriously said to babies and most preschoolers— or our pets. We don't seriously expect our dogs and cats to appraise a novel situation, like a fish falling out of the refrigerator, with an eye toward saving it for the dinner guests tonight.

An ability to guess the consequences of a course of action is the foundation of ethics. Free will implies not only the choice between known alternatives, but an ability to imagine novel alternatives and to shape them up into something of quality. Many animals use trial and error, but we humans do a great deal of it "off line," before actually acting in the real world. The process of contemplation and mental rehearsal that shapes up novel variations would appear to lie at the core of some of our most cherished human attributes. *How* do we do that?

Creating novelty isn't difficult. New arrangements of old things will do.

Everyone thinks that mutations (as when a cosmic ray comes along and knocks a DNA base out of position, allowing another to fill in) are where novel genes come from. Nature actually has two other mechanisms that are more important: copying errors and shuffling the deck. Anyone with a disk drive knows that copying errors is the way things are, that procedures had to be invented to detect them (such as those pesky check sums) and correct them (such as the error-correcting codes which are now commonplace). All that's required to achieve novelty is to relax vigilance.

But nature occasionally works hard at mixing up things, each time that a sperm or ovum is made; the genes on both chromosomes of a pair are shuffled (what's known as crossing over during meiosis) before being segregated into the new chromosome arrangement of the sperm or ovum. And, of course, fertilization of an ovum by another individual's sperm creates a new third individual, one that has a choice (in most cases) between using a gene inherited from the mother or the version of it inherited from the father.

Quality is the big problem, not novelty as such. Nature's usual approach to quality is to try lots of things and see what works, letting the others fall by the wayside. For example, lots of sperm are defective, missing essential chromosomes; should they fertilize an ovum, development will fail at some point, usually so early that pregnancy isn't noticed. For this and other reasons, over 80 percent of human conceptions fail, most in the first six weeks (this spontaneous abortion rate is far more significant than even the highest rates of induced abortions).

There are often high rates of infant and juvenile mortality as well; only a few individuals of any species survive long enough to become sexually mature and themselves become parents. As Charles Darwin first realized in about 1838, this is a way that plants and animals change over many generations into versions that are better suited to environmental circumstances. Nature throws up a lot of variations with each new generation, and some are better suited to the environment than others. Eventually, a form of quality emerges through this shaping-up process.

When Darwin explained how evolution might produce

more and more complex animals, it started the psychologists thinking about thought itself. Might the mind work the same way as Darwin's mechanism for shaping a new species? Might a new thought be shaped by a similar process of variation and selection?

Most random variations on a standard behavior, even if only changing the order of actions, are less efficient, and some are dangerous ("look after you leap"). Again, quality is the problem, not novelty per se. Most animals confine themselves to well-tested solutions inherited from ancestors that survived long enough to successfully reproduce. New combinations are sometimes tried out in play as juveniles, but adults are far less playful.

By 1880, in an article in the *Atlantic Monthly,* the pioneering American psychologist William James (who invented the literary term "stream of consciousness") had the basic idea:

> [T]he new conceptions, emotions, and active tendencies which evolve are originally *produced* in the shape of random images, fancies, accidental outbursts of spontaneous variations in the functional activity of the excessively unstable human brain, which the outer environment simply confirms or refutes, preserves or destroys—selects, in short, just as it selects morphological and social variations due to molecular accidents of an analogous sort.

His French contemporary, Paul Souriau, writing in 1881, said much the same thing:

> We know how the series of our thoughts must end, but . . . it is evident that there is no way to begin except at random. Our mind takes up the first path that it finds open before

it, perceives that it is a false route, retraces its steps and takes another direction. . . . By a kind of artificial selection, we can . . . substantially perfect our own thought and make it more and more logical.

James and Souriau were building on the even more basic idea of Alexander Bain, concerning trial and error. Writing in Scotland in 1855, Bain initially employed the phrase *trial and error* when considering the mastery of motor skills such as swimming. Through persistent effort, Bain said, the swimmer stumbles upon the "happy combination" of required movements and can then proceed to practice them. He suggested that the swimmer needed a sense of the effect to be produced, a command of the elements, and that he then used trial and error until the desired effect was actually produced. This is what a Darwinian process can use to shape up a thought—which is, after all, a plan for a movement, such as what to say next.

Surprisingly, no one seemed to know what to do next, to make the link between the basic idea of Darwinian thought and the rest of psychology and neurobiology. For more than a century, this key idea has lain around like a seed in poor soil, trying to take hold. One problem is that it is easy (even for scientists) to adopt a cartoon version of Darwinism— survival of the fittest, or selective survival—and fail to appreciate the rest of the process.

The basic Darwinian idea is deceptively simple. Animals always reproduce, but all of their offspring don't manage to grow up to have babies themselves—they overproduce. There is a lot of variation in the offspring of the same two parents; each offspring (identical twins and clones excepted) gets a different set of shuffled chromosomes.

Operating on this generated diversity is selective survival. Some variants survive into adulthood better than others, and so, the next generation's variations are based on the survivor's genes. Some are better, most are worse, but they center around an advanced position because the worst ones tend to die young. And the next generation is, for the average survivor into adulthood, even better suited to the environment's particular collection of food, climate, predators, nesting sites, etc.

We usually think in terms of millennia as the time scale of this process that can evolve a new species. With artificial selection by animal breeders, substantial effects can be produced in a dozen generations. But the process can operate on the time scale of the immune response, as new antibodies are shaped up by success in killing off invading molecules; within a week or two, antibody shapes can be evolved that have a key-and-lock specificity for a foreign molecule. Might the same process suffice for the time scale of thought and action?

It is worth restating the six essentials of a Darwinian process a bit more abstractly, so we can separate the principles from the particulars:

- There is a pattern involved (typically, a string of DNA bases—but the pattern could also be a musical melody or the brain pattern associated with a thought).
- Copies are somehow made of this pattern (as when cells divide, but also when someone whistles a tune he's heard).
- Variations on the pattern sometimes occur, whether by copying errors or by shuffling the deck.
- Variant patterns compete for occupation of a limited space (as when bluegrass and crabgrass compete for your backyard).
- The relative success of the variant patterns is influenced by a multifaceted environment (for grass, it's hours of sunlight, soil nutrients, how often it's watered, how often it's cut, etc.).

- And, most important, the process has a loop. The next generation is based on which variants survived to maturity, and that shifts the base from which the surviving variants spread their own reproductive bets. And the next, and the next. This differential survival means that the variation process is not truly random. Instead, it is based on those patterns that have survived the multifaceted environment's selection process. A spread around the currently successful is created; most will be worse, but some may be better.

Not every process that makes copies of patterns is going to qualify as Darwinian. Photocopy and fax machines make copies of the ink patterns on a sheet of paper, but there is usually no loop.

If you do make copies of copies, for dozens of generations, you will see some copying errors (especially if copying gray-scale photographs). Now you've satisfied the first three conditions—but you still don't have competition for a work space that is biased by a multifaceted environment, nor an advantage in reproduction for certain variants.

Similarly, you can have selective survival without the rest of the Darwinian process. You will find more fifteen-year-old Volvos still on the road than you will Fiats of the same age. But the fifteen-year-old Volvo doesn't reproduce. Nor do brain cells—though the connections between them (synapses) are edited over time. In the brain of an infant, there are many connections between nerve cells that don't survive into adulthood. But this selective survival (random connections that prove to be useful) isn't proper Darwinism either, unless the surviving connection patterns somehow manage to reproduce themselves elsewhere in that brain (or perhaps through mimicry in someone else's brain). And, even if they did, this pattern copying would still have to satisfy the requirement for a loop where reproduction with new variation is biased toward the more successful.

Selective survival is a powerful mechanism that produces crystals in nonliving matter as well as economic patterns in cultural evolution. Selective survival is a problem for all business enterprises, especially small ones, but it leads—at least in capitalist free-market theories—to a better fit with "what works."

Selective survival of all sorts is sometimes called Darwinian (Darwin was annoyed when Herbert Spencer started talking of social Darwinism). But selective survival per se can even be seen in nonliving systems, as when flowing water carries away the sand grains and leaves the pebbles behind on a beach.

Full-fledged Darwinism is even more powerful, but it requires differential reproduction of the more successful. Economics has some recent examples in fast-food franchises, where copies are produced of the more successful of an earlier generation. Indeed, they seem to be in a competition with their variants for a limited "work space." If they close the loop by generating new variations on the more successful (imagine a MacUpscale and a MacEconomy splitting off from one of the chains), they may provide another example of a Darwinian process evolving new complexity.

When people call something "Darwinian," they're usually referring to only part of the Darwinian process, something that uses only several of the six essentials. And, so powerful are the words we use, this overly loose terminology has meant that people (scientists included) haven't realized what was left out.

Indeed, the second reason why the Darwinian thought idea wasn't fleshed out earlier is that it has taken a while to realize that thought patterns might need to be copied—and that the copies might need to compete with copies of alternative thoughts. Since we haven't known how to describe the neural activities underlying thought, we haven't been able to think about copying. But copying is a major clue about what the thought process must be like; it's a constraint that considerably reduces the possibilities.

In the early 1950s, during the search for the genetic code, molecular biologists were acutely aware of the need for a molecular process that could somehow make copies of itself during cell division. The reason why the double helix structure was so satisfying in 1953 was that it solved the copying problem. In subsequent years, the genetic code (the translation table between DNA triplets and amino acid strings) was worked out. Perhaps we too can identify the cerebral code that represents an object or idea, with the aid of looking at what cerebral patterns can be semiaccurately copied.

Thoughts are just combinations of sensations and memories—or, looked at another way, thoughts are movements that haven't happened yet (and maybe never will). The brain produces movements with a barrage of nerve impulses going to the muscles, whether limbs or larynx. But what determines the details of this barrage?

Sometimes, it is simply an innate rhythm such as the ones which produce chewing, breathing, and walking. Sometimes there is time for lots of corrections, as when you lift a coffee cup and discover that it weighs less than you thought; before it hits your nose, you manage to make some corrections to your arm muscles. But some movements are so quick (over and done in one eighth of a second) that no feedback is possible: throwing, hammering, clubbing, kicking, spitting (including "spitting out a word"). We call these ballistic movements; they're particularly interesting because they require that a complete plan be evolved before acting. During "get set," you have to produce the perfect plan. A plan for a movement is like the roll for a player piano: eighty-eight output channels, one for each key, and the times at which each key is struck. To hammer or throw indeed requires coordinating close to eighty-eight muscles, so think of a sheet of music as a plan for a spatiotemporal pattern—all of those chords, melodies, and interweaving patterns we call musical.

In 1949, the Canadian psychologist Donald Hebb formulated his cell-assembly hypothesis, stating that evoking a memory required reconstituting a pattern of activity in a whole group of neurons. We now think of Hebb's cell assembly more generally as a spatiotemporal pattern in the brain which represents an object, an action, or an abstraction such as an idea. Each is like a musical melody and, I calculate, takes up about as much space in the brain as would the head of a pin (just imagine that the pinhead is hexagonal in shape).

Memories are mere spatial patterns frozen in time—that sheet of music waiting for a pianist, or the ruts in a washboarded road, lying in wait for something to come along and interact with them to produce a spatiotemporal pattern in the form of live music or a bouncing car. A Darwinian model of mind suggests that an activated memory can interact with other plans for action, compete for occupation of a work space. A passive memory, like those ruts in the road, can also serve as an aspect of the environment that biases a competition—in short, both the current real-time environment and memories of past environments can bias a competition that shapes up a thought.

So we have a pattern—that musiclike thought in the brain—and we have selective survival biased by a multifaceted environment. How can thoughts be copied to produce dozens of identical pinheads? How can their variants compete for a work space, the same as bluegrass and crabgrass compete for a backyard? How can the loop be closed?

All of the currently active cerebral codes in the brain, whether for objects like apples or for skilled finger movements such as dialing a telephone, are thought to be spatiotemporal patterns. To move a code from one part of the brain to another, it isn't physically sent, as a letter is mailed. Rather, it has to be copied much like a fax machine makes a copy of the pattern on one sheet of paper onto a new sheet of paper at the remote location. The transmission of a neural code involves making a copy of a spatiotemporal pattern, sometimes a distant copy via the fibers of the *corpus callosum* but often a nearby copy, much in the manner that a crystal grows.

The cerebral cortex of the brain, which is where thoughts are most likely to arise, has circuitry for copying spatiotemporal patterns in an immediately adjacent region less than a millimeter away. All primates have this wiring, but it is not known how often they use it. The cerebral cortex is a big sheet—if peeled off and flattened out, it would be about the size of enough pie crust to cover four pies. There are at least 104 standard subdivisions. While some areas of cortex might be committed to full-time specialization, other areas might often support sideways copying and be erasable work space for Darwinian shaping-up processes.

The picture that emerges from theoretical considerations is one of a patchwork quilt, some parts of which enlarge at the expense of their neighbors as one code comes to dominate. As you try to decide whether to pick an apple or an orange from the fruit bowl on the table, the cerebral code for *apple* may be having a copying competition with the one for *orange.* When one code has enough active copies to trip the action circuits, you reach for the apple. But the orange codes aren't entirely banished; they could linger in the background as subconscious thoughts.

When you try to remember someone's name, but without initial success, the candidate codes might continue copying with variations for the next half hour until, suddenly, Jane Smith's name seems to "pop into your mind." Our conscious thought may be only the currently dominant pattern in the copying competition, with many other variants competing for dominance (just as the bluegrass competes with the crabgrass for my backyard), one of which will win a moment later when your thoughts seem to shift focus.

The Darwinian process is something of a default mechanism when there is lots of copying going on, and so we might expect a busy brain to use it. Perhaps human thought is more complicated than this, with shortcuts so completely dominating the picture that the Darwinian aspects are minor. Certainly, the language mechanisms in our brain must involve a lot of rule-based shortcuts, judging from how children make relatively sudden transitions from speaking simple sentences to speaking much more complicated ones, during their third year of life. It may be that the Darwinian processes are only the frosting on the cake, that much is routine and rule-bound.

But the frosting isn't just writing poetry or creating scientific theories (such as this one). We often deal with novel situations in creative ways, as when deciding what to fix for dinner tonight. We survey what's already in the refrigerator and on the kitchen shelves. We think about a few alternatives, keeping track of what else we might have to fetch from the grocery store. And we sometimes combine these elements into a stew, or a combination of dishes that we've never had before. All of this can flash though the mind within seconds—and that's probably a Darwinian process at work, as is speculating about what tomorrow might bring.

WILLIAM H. CALVIN is a theoretical neurophysiologist on the faculty of the University of Washington School of Medicine. Rather than teaching students to support his research, he now stays off the payroll and writes books for general readers. *The Throwing Madonna, The Cerebral Symphony,* and *The Ascent of Mind* are about brains and evolution; *The River That Flows Uphill* is about his two-week trip down the rapids of the Colorado River in the bottom of the Grand Canyon; *How the Shaman Stole the Moon,* set in the canyonlands of the Southwest, is his "hobby book" concerning pre-Stonehenge entry-level methods for predicting eclipses. In *Conversations with Neil's Brain: The Neural Nature of Thought and Language,* coauthored with neurosurgeon George Ojemann, Calvin narrates a day of neurosurgery for epilepsy, focusing on how an internal voice is generated, one that occasionally speaks aloud. Calvin occasionally strays outside evolution and neuroscience, as in his articles for the *Whole Earth Review* on abrupt climate change, foretelling eclipses, and his commentary on Verner Vinge's article about technological singularities.

The Puzzle of Averages

Michael S. Gazzaniga

Numbers, massaged with statistics, become facts. Numbers without statistics become anecdotes. In the course of scientific training, it is a noble enterprise to teach the young investigator that one simple observation may not reveal the true pattern or processes of mother nature. Many observations are needed, and the data must be averaged and tested for significance. The hoary presence of chance phenomena is always a possibility, and as science aims to distinguish between random events and true processes, the scientist must always be on guard against chance occurrences or random effects.

But strange things begin to happen when we seek averages for all data while striving to make our observations more powerful. Nowhere does the problem become more nonsensical than in the area of labor statistics. Studies have shown that in year X, there were so many people living in town A. In year Y, town B boomed and town A declined. The usual inference made is that the workforce shifted from town A to town B. Yet, when individual citizens are actually studied, very few went to town B from town A. True, they moved out of town A, but moved to towns C, D, and E. Talk about the "average" is particularly dispiriting in the social sciences from which we learn what the average man thinks, or feels, or whatever. The mentality of the averages is everywhere. Have you ever wondered what it means that each American family has 2.3 children? I suppose that measure helps demographers and city planners, but it does not indicate what is going on in personal lives. Living by averages is scary.

Recently, George Miller, the doyen of American psychology, told me about a study a colleague of his, Joyce Weil, had carried out many years ago on language development in children. There was a story told in those days that children went through some kind of orderly progression in their linguistic development, such that four-year-olds did things three-year-olds didn't do, and three-year-olds, in turn, did things that two-year-olds didn't do. Children were studied at these ages and their language capacities were studied and their patterns averaged. When the data were averaged in this way, an orderly pattern emerged and it was then seized upon by psychologists for support of their orderly theories about language development. But Weil actually looked at what individual kids did in the course of their development. This longitudinal approach revealed that individual kids didn't follow the average pattern at all. The average pattern and the theory developed from only considering that the averaged data were a statistical anomaly that reflected nothing in the real world. Weil couldn't get the paper published. The scientific community's belief in averages had produced a theory which was now entrenched in the scientific community.

I bring this up because I have come upon a simple truth

as well, when it comes to the issue of averaging. It has to do with the need, as always, to try to find relationships in an otherwise noisy set of data. In this case, we were looking for unique areas in the two hemispheres of the brain that may show asymmetry. Over a hundred years of clinical observations indicate that the left hemisphere of the human brain is specialized for cognitive functions, in particular, language. In hunting for that specific part of the left brain, several scientists had suggested that a part of the temporal lobe is bigger on the left side than on the right. But we now know that this much-heralded asymmetry reflects only the folding pattern of the cortex and not its actual surface area. Since the original anatomical studies were done in the mid-sixties, there have been major advances in computer-based brain imaging. These new methods now allow us to take the picture of living brains and to analyze their three-dimensional structure in detail, a process that corrects for the errors inherent in doing a two-dimensional analysis. When normal human brains are analyzed in this way, the temporal lobe asymmetry goes away.

But this same technology also allows us to divide up each half brain into some twenty-seven regions of interest and then to compare the left and the right brain and to look for other possible areas that might be asymmetrical. Of course, we averaged the surface-area measurements for each region and then compared left and right brains. Within the limits of what we chose to call asymmetrical, we discovered that on the average, there are no human brain asymmetries. The problem is, however, that each individual brain we examined had a unique profile. Some were leftward asymmetrical and some were rightward. Some have larger asymmetrical areas in the frontal lobes, others in the occipital lobe, and still others in the parietal and temporal lobes.

Averages, in this instance, construct a picture about some kind of idealized brain. In the real biological world, brains are obviously very asymmetrical and the variation is great. When all of these differences are averaged out, we construct a brain that doesn't exist in nature. Each brain has a different pattern of asymmetry and in that fact lies the truth that any simple claim about a particular asymmetrical region being particularly involved in a particular mental function is doomed to be wrong. In short, averages can be artifactual and, as we have just seen, can even be systematically false. Indeed, it might even be that the unique patterns of our own brains might just be responsible for our own unique minds.

MICHAEL S. GAZZANIGA is director of the Center for Neuroscience and professor of neurology and psychology at the University of California, Davis. He has spent the past thirty years gaining a deeper understanding of split-brain patients. His research in this area began in 1960 working with Richard Sperry, who won the Nobel Prize and who, along with Ronald Myers, discovered the split-brain phenomenon—that when the fibers connecting the brain's hemispheres were severed, information visually trained to one half-brain did not transfer to the other half-brain. He is the author of *The Social Brain; Nature's Mind;* and *Mind Matters.*

Ceteris Paribus (All Else Being Equal)

Pascal Boyer

To a student who told me that some theories I had explained (not mine) were "too simplistic," and was shocked when I replied that this was their main virtue.

Dear L.,

Good thinking requires intellectual style. Style consists mainly in avoiding some crucial mistakes, which unfortunately are as widespread as they are damaging. Thinking requires that you use the appropriate tools, and the tool I want to tell you about is perhaps the most modest of all, a tool so discrete in most theories and arguments that some people do not even notice its existence. Yet that tool is used; not only that, if it is not used, then you simply cannot do science. And in my view, you cannot even think about any problem at all, scientific or otherwise, in a way that makes genuine sense. Let me go even further. There is a distinction between those who use this tool and those who do not; it is always difficult for these two kinds of people to communicate. The tool has a Latin name which makes it respectable and mysterious, *ceteris paribus*, and an English translation which is, in fact, more deceptive for being so simple, *all else* (or *other things*) *being equal*. If you spend some time pondering what this phrase really means and how it can be used, you will soon touch upon some very important aspects of intellectual style.

In order to explain the importance of this tool, let me take an example that is only partly imaginary. Let us suppose you have been wondering about the way people make decisions in situations where they cannot have access to all the relevant information. People buy lottery tickets without knowing which numbers will be drawn; they apply for jobs without knowing whether they are the best candidate; they behave in a friendly way toward people whom it might be better to avoid. In such situations, you know that people are likely to make a certain decision if they think that it might bring about the result they expect, and if the expected result is one they desire. You have noticed, and this is your great discovery, that the likelihood that people will make a decision can be predicted by multiplying these two factors. That is to say, if you can measure the probability of success and the desirability of the outcome, then the product of those two measurements will give you the probability that the subject will take that course of action. (We'll leave aside the technicalities involved, which you have solved, since you have a keen mathematical mind.) This principle has interesting consequences. For instance, it seems to predict that people are *equally* likely to make a certain decision, in two very different situations: (a) when the outcome is highly desirable, although not very likely to occur, and (b) when the outcome is not quite as desirable, but slightly more likely to be achieved. This is a consequence of your idea of a multiplication: A product is the same when one factor is increased and the other decreased in the same proportion. You feel quite good about your discovery, because it seems to apply reasonably well to many different situations. You can explain why some lotteries are more popular than others in that way; you can also explain why people apply for jobs that they have only a slim chance of getting but which seem so desirable.

Let us suppose that you present your discovery to various audiences. In this imaginary situation, no one has yet thought about decision-making and uncertainty in any formal way, so that your "principle" is something quite new. I can predict that you will get two kinds of reactions, leaving aside people who simply do not understand what you are talking about. First, some people will make all sorts of objections about various aspects of your theory. For instance, they will say that you should not talk about the probability of success or the desirability of an outcome, but about what the subject *perceives* of those aspects. People do not act on the basis of how things really are, but on the basis of how they think they are. Your audience will also tell you that your principle would be more realistic if you added a third factor to the equation, namely the cost involved in making a decision. Given an equal chance of winning an equal amount, people will be more likely to buy the lottery ticket if it is cheaper. There will be many other objections of this kind. All this may be a bit painful to hear, as it shows that your theory was far from perfect. But you are a genuine scientist, and the pursuit of truth matters more to you than mere vanity. Back to the drawing board, then.

Unfortunately, another kind of criticism is likely to be made against your principle. Some people will object, not to the way you have constructed your theory, but to the very idea of such a theory. They will tell you this:

"It makes no sense to talk about people making decisions in the abstract, without taking the context into account. First, the decisions are made in different situations. Buying a lottery ticket is not at all the same as applying for a job, so why should there be a common "logic" to those different situations? Surely, what they do with lottery tickets depends on their ideas about lotteries, and what they do with job applications depends on their ambitions, and you must study these specific contexts in detail. Second, we all know people who act irrationally. They may be drunk when they buy their lottery ticket; they may have unrealistic notions of their own capacities; and in many cases, they are moved by irrational urges and passions. So your principle does not make any sense."

This time, too, you are hurt that people could fail to be impressed by your theory—but in this imaginary exchange you are right and they are wrong, and I think it is important to understand why this is so. Once you have got over your anger, you will probably think that they have just missed the point of your theory. And this might be your stinging reply:

"First, my ambition was not to describe everything about gambling or job applications, but to describe their common features. If you describe what two situations have in common, it is not because you think they have everything in common, but because you think that the aspects they have in common might be important. Second, I was trying to describe what a reasonable person tends to do, when he or she tries to bring about desirable results and avoid undesirable ones. I chose to ignore fairly exceptional cases, where people are perversely trying to fail or are mentally disturbed. So how could you raise such irrelevant objections?"

They could, because they have not understood that your principle, when you formulated it, was prefaced with the phrase *all else being equal.* Perhaps you did not realize that yourself. They could raise irrelevant points because they do not understand what this phrase means, or because they do not like the intellectual style it represents. For many people, doing science consists in discovering "what really happens," beyond prejudice and received wisdom. Scientists are seen as people who describe things the way they really are. So it seems that one's theories should always be "true to life." This is all very flattering to scientists, but it is also complete nonsense. Scientific theories are "true to life" only in the sense that *evidence* is the only tribunal that judges right and wrong in science. An embarrassing, unexplained fact carries more weight than a satisfactory, elegant theory, and that is what makes scientific activities so frustrating sometimes. In another sense, scientific ideas are not, cannot be, and *should not* be "true to life." And this is all that the phrase *all else being equal* really means. There is nothing terribly complicated about this phrase. In fact, in the paragraph above, you just explained what it means, in your reply to those silly objections. To put it in more abstract terms: Producing a theory does not mean taking into account all possible aspects of the phenomena you describe. On the contrary, it means that you focus on some aspects that can be described in terms of abstract generalizations, assuming, for the sake of simplicity, that all other aspects are "neutralized," i.e., that they are "equal." This is where some people will get flustered, and tell you that all other things are *not* equal, that they vary from situation to situation. How can you just *decide* not to consider what actually happens, in all its richness and complexity?

Whatever phenomena you want to explain, you must remember that all sound theories are based on that decision to consider all other things equal. Without it, you cannot describe even the simplest phenomena. In fact, we have illustrations of this in our everyday environment. Let me take a simple example, which does not involve advanced science. When your eyes are tested, your visual acuity is generally measured by asking you to read the letters printed on a screen, about ten feet in front of your chair. These are isolated capital letters, projected on a bright panel; the rest of the room is in a semidarkness that highlights the panel. Doctors seem content to measure your eyesight, and prescribe lenses, on the basis of your performance in this kind of test. If you do not know or understand the *ceteris paribus* principle, you might object to such assurance, on the grounds that the test does not "take into account" the natural contexts of ordinary vision. What we usually read are meaningful words and sentences, not isolated capital letters; we seldom read on a bright screen in a dark room; and in most cases, objects display a color and a texture and some contiguous connections with other objects. So what the doctor measured cannot be your "real" visual acuity.

Let me return to your scientific theory of decision-making. With your "principle" (which I think is extremely simple and has the advantage of being true), you have just started a brilliant career as a specialist of decision-making under uncertainty. You should not be unduly disturbed by irrelevant objections. You must realize, however, that your path will be strewn with such absurd arguments. This is because, wherever you go, you will meet people who do not understand the *ceteris paribus* as often as people who do. They have different frames of mind, and belong to two different cultures. This is very much what the Russian writer Alexander Zinoviev had in mind when he wrote in his *Yawning Heights* of the "two principles" that govern intellectual activity: "The scientific principle produces abstractions, the anti-scientific principle destroys them on the grounds that such and such has not been considered. The scientific principle establishes strict concepts, the anti-scientific principle makes them ambiguous on the pretext of thus revealing their true variety."

But again, why should we accept this intellectual program? The main reason for thinking in that way, for isolating different factors and idealizing away from reality, is that this is what you need to do in order to do science, and that science has been more successful than any other intellectual enterprise so far. It has explained more than any other way of thinking about the world. So it is just plain *better* . . . all else being equal, of course.

Pascal Boyer

PASCAL BOYER is a senior research fellow of King's College, Cambridge. His main research is at the interface of anthropology and cognitive science, showing how various cultural phenomena are constrained by universal characteristics of human minds. In the past five years, he has turned his interest to religious representations, and is engaged in experimental cognitive work on children and adults. The point of this research is to demonstrate that human minds are particularly susceptible to entertain representations that violate their intuitive, tacit understandings of the natural world. Religious ideas are not answers to universal human questions, nor do they provide explanations for complex experiences. Rather, they are a side effect of the kind of cognitive capacities evolution has given us. The anthropological aspects of the research are presented in his books *Tradition as Truth and Communication* and *The Naturalness of Religious Ideas.*

On Taking Another Look

Nicholas Humphrey

"How often have I said to you," Sherlock Holmes observed to Dr. Watson, "that when you have eliminated the impossible, whatever remains, however improbable, must be the truth?" And how often do we need to be reminded that this is a maxim that is quite generally ignored by human beings?

Here is an immediate test to prove the point. Figure 1 shows a photograph of a strange object, created some years ago in the laboratory of Professor Richard Gregory. What do you see this as being a picture of? What explanation does your mind construct for the data arriving at your eyes?

You see it, presumably, as a picture of the so-called impossible triangle: that is, as a picture of a solid triangular object whose parts work perfectly in isolation from one another but whose whole refuses to add up—an object that could not possibly exist in ordinary three-dimensional space.

Figure 1

Yet the fact is that the object in the picture does exist in ordinary space. The picture is based on an unretouched photograph of a real object, taken from life, with no kind of optical trickery involved. Indeed, if you were to have been positioned where the camera was at the moment the shutter clicked, you would have seen the real object exactly as you are seeing it on the page.

What, then, should be your attitude to this apparent paradox? Should you perhaps (with an open mind, trusting your personal experience) believe what you unquestionably see, accept that what you always thought could not exist actually does exist, and abandon your long-standing assumptions about the structure of the "normal" world? Or, taking heed of Holmes's dictum, would you do better instead to make a principled stand against impossibility and go in search of the improbable?

The answer, of course, is that you should do the second. For the fact is that Gregory, far from creating some kind of "paranormal" object that defies the rules of 3-d space, has merely created a perfectly normal object that defies the rules of human expectation. The true shape of Gregory's "improbable triangle" is revealed from another camera position in Figure 2.

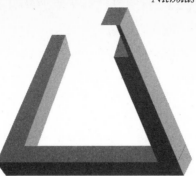

Figure 2

It is, as it turns out, a most unusual object (there may be only a couple of such objects in existence in the universe). And it has been photographed for Figure 1 from a most unusual point of view (to get this first picture, the camera has had to be placed at the one-and-only position from which the object looks like this). But there it is. And now that you have seen the true solution, presumably you will no longer be taken in.

If only it were so! You look at Figure 2. And now you look back at Figure 1. What do you see this time around? Almost certainly, you still see exactly what you saw before: the impossibility rather than the improbability! Even when prompted in the right direction, you happily, almost casually, continue to "make sense" of the data in a nonsensical way. Your mind, it seems, cannot help choosing the attractively simple—even if mad—interpretation over the unattractively complicated—even if sane—one. Logic and common sense are being made to play second fiddle to a perceptual ideal of wholeness and completion.

There are many examples in the wider world of human politics and culture where something similar happens, that is to say, where common sense gets overridden by some kind of seductively simple explanatory principle—ethical, political, religious, or even scientific. For, if there is one thing that human beings are amazingly prone to (perhaps we might say good at), it is in emulating the camera operator who took the photograph of Figure 1 and maneuvering themselves into just the one ideological position from which an impossible, even absurd explanation of the "facts of life" happens to look attractively simple and robust.

This special position may be called, for example, Christianity, or Marxism, or Nationalism, or Psychoanalysis—maybe even some forms of science, or scientism. It may be an ideological position that appeals only to some of the human population some of the time or one that appeals to all of the population all of the time. But, whichever it is, to those people who, in relation to a particular problem, are currently emplaced in this position, this will almost certainly seem to be the only reasonable place there is to be. "Here I stand," in the words of Martin Luther, "I can do no other"; and the absolute rightness of the stance will seem to be confirmed by the very fact that it permits the welcome solution to the problem that it does.

Yet the telltale sign of what is happening will always be that the solution works only from this one position—and that if the observer were able to shift perspective, even slightly, the gaps in the explanation would appear. Of course, the trick—for those who want to keep faith and save appearances—is not to shift position, or to pull rapidly back if ever so tempted.

The lesson is that when would-be gurus offer us final answers to any of life's puzzles, a way of looking at things that brings everything together, the last word on "How Things Are"—we should be watchful. By all means, let us say: "Thank you, it makes a pretty picture." But we should always be prepared to take another look.

NICHOLAS HUMPHREY is a theoretical psychologist who has held research and teaching posts at both Oxford and Cambridge, as well as fellowships in the United States and in Germany. His books include *Consciousness Regained* and *The Inner Eye,* as well as *A History of the Mind.* His interests are wide ranging: He studied mountain gorillas with Dian Fossey in Rwanda; made important discoveries about the brain mechanisms underlying vision; proposed the now-celebrated theory of the "social function of human intellect"; and is the only scientist ever to edit the literary journal *Granta.* He has been the recipient of several honors, including the Martin Luther King, Jr., Memorial Prize in 1985.

What to Know, How to Learn It

Roger C. Schank

W hat should an educated person know? In school, little time is devoted to answering this question. The school already knows what you need to know: vocabulary about phyla, the plots of various pieces of last century's literature, and how to prove a theorem about triangles. When you try to get computers to know things in order to make them act "intelligently," however, it turns out that these are not at all the sorts of things they need to know. Computers need to know how to do things, how to comprehend what others have done and said, and how to learn from the mistakes it makes in doing all this.

To educate a computer means giving it the ability to make inferences (if John hits Mary, it means she is probably hurt, angry, and may hit him); to infer beliefs (if the United States bombs Iraq, America must believe that violence is justified by the need to cater to one's economic interests); and to learn from failure (when you order filet mignon on an airplane and it is overcooked mush, you might want to remember this the next time you're ready to board a plane—and get a yogurt first). These are exactly the same things people must learn how to do. We know that a computer, for instance, can be programmed to have encyclopedic knowledge about random facts, but I don't believe that this would mean it is "intelligent," anymore than I would consider a person who merely has the ability to spout random facts to be intelligent. But despite what we know about how people learn and the very makeup of intelligence, schools remain firmly grounded in the learning model that emphasizes facts and downplays doing. This separation of learning from doing is very detrimental to all.

It has become fashionable recently to define intelligence by using various "literacy lists." The bookstores are full of lists of different kinds of facts—scientific, cultural, even religious facts—all purporting to explain exactly what it is that a person must know to be "literate." The idea here is that being educated means knowing stuff. Implicit in all this is that we have, as a society, agreed on what stuff everyone should know, and decided that information delivery is the role of education.

Do not believe it. There is no set of stuff that everyone should know. What? No George Washington? No Gettysburg Address? It doesn't hurt to know these things, of course. But it does hurt to adopt the position that since one should know these things, teaching them to students is what learning is all about. This makes school a fairly boring, stressful, and irrelevant place, as you may have already discovered.

Facts are not the currency of learning, nor does mastery of them indicate anything about an educated person. Facts play a big role in the education system because they are so easy to test. And it is tests (usually highly irrelevant tests) that have helped shape your learning since you were six. Curiously, most important things that people know they cannot explicitly recall or state as facts. What is the right way to get the person of your dreams interested in you? How does one pursue a successful career? Was the United States wrong to believe in Manifest Destiny? Is the situation in Bosnia really all that similar to Nazi Germany, or is it more like Vietnam? An educated person might have answers for these questions. But they are not simple questions and there are no simple answers for them. Being educated means being able to understand the questions and knowing enough relevant history to be able to make reasoned arguments. Making reasoned arguments, not citing history, is the key issue here. Learning to think and express what one has thought in a persuasive way is the real stuff of education.

What is the currency of learning? It is a preparedness to be wrong, a willingness to fail, and the ability to focus on one's confusion in hope of being able to create or being able to understand an explanation that will make things clearer.

For this reason, the way that stuff is imparted is far more important than the stuff itself. It isn't what you know but how you come to know it that matters. Typically, what we learn in school comes to us through the process of memorization. But memorizing something doesn't mean being able to remember it later when that information might be useful. Information acquired in one context cannot be readily used in another context. While it is not easy to resist a school's attempts to force you to memorize facts, it is important to recognize that merely memorizing things doesn't mean you'll know much. Being able to articulate facts is useful for passing tests, impressing your friends, and doing well on quiz shows, but for little else.

We learn best what we want to learn—information which helps us accomplish goals we have set for ourselves. Computers—intelligent computers—can help us do this by providing safe but exciting environments in which to learn. Such computers can be "taught" to ask questions, provide helpful information, and be endlessly patient as a user tries to solve problems.

For the Museum of Science and Industry in Chicago, the Institute for the Learning Sciences built a computer-based exhibit which shows how this process can work. The museum wanted to teach visitors about sickle-cell disease; "Sickle-Cell Counselor" does just that by allowing visitors to play the role of genetic counselors. Of course the visitors have no intention of becoming genetic counselors. But by presenting them with a challenging problem, their natural motivation to learn is captured.

The problem posed by the program is to advise couples worried about having children because they suspect their children may be at risk for inheriting the sickle-cell gene. Sickle-Cell Counselor provides simulations of actions genetic counselors actually take with their clients, and provides access for the user to human experts (via videotape) who help solve the problems at hand.

Sickle-Cell Counselor continues to be a popular exhibit at the museum, where visitors often spend up to half an hour exploring its varied opportunities for learning. This is much more time than museum visitors generally spend at an exhibit, but they stay because the learning experience offered is a realistic one, providing goals they become interested in accomplishing.

Not long ago, I asked students in an undergraduate class what they had learned recently. They recited facts they had learned in other classes, but they had no idea when they would ever use this information again. When I asked a class of graduate students the same question, they only told me things they had learned about living. They had just rented apartments for the first time, so I heard a lot about how to cook and clean, but they also talked about what they had learned in school that had been of use in projects they were now trying to complete. Graduate students typically focus on accomplishing tasks. They learn what they need that will help them with those tasks. For them, learning means acquiring knowledge in service of a goal. But unless you consider passing tests a goal, you won't find this pattern repeated very much before graduate school.

To make computers intelligent, we must teach them to direct their own learning. We can't just pour facts into computer memories, because they won't know what to do with the knowledge they've gained. But if they acquire knowledge in the course of doing something, then the placement of that knowledge in their memory is easy; it gets placed at the point where it was learned so it can modify the procedures that were wrong when that new knowledge was acquired.

You, too, must learn to direct your own learning. Context provides structure for learning, so putting yourself in many different situations, or trying many different things, is very important. Not knowing how to do something in a situation causes a person to focus on learning how to accomplish the unknown, so keep trying to do new things, and attempt to understand why you fail and why you succeed. Don't allow people to lecture at you, imparting information you don't want. On the other hand, you must demand to be taught after you have struggled to accomplish something and have had difficulties.

The key to learning, after you've tried to do something,

failed, and availed yourself of needed help, lies in the process of generalization. It is not sufficient just to learn how to function in a given situation; you must also learn to generalize what you have learned so that it applies elsewhere. If you cannot do this, you will have gained a narrow collection of unrelated expertise, useful only for individual domains, but useless otherwise.

It is not possible to make generalizations after acquiring information without consciously attempting to do so. However, attempting to do so involves coming up with generalizations that are inherently untestable, that remain only hypotheses, and that are not in themselves facts. Even so, do not be afraid to test out your generalizations on people you know. They will undoubtedly tell you that you are wrong, but ask them to support their assertions. In general, people are afraid of new generalizations because they cannot know if they are right or not. People fear the unknown, but real learning and real insight depend upon an examination of the unknown and perhaps the unknowable. Propose new generalizations and be prepared to defend them. Ignore new facts that were presented to you unless they were presented in order to help you do something—answer a question, or alter one of your generalizations.

To see why facts don't matter, consider the value of a typical fact that most literacy lists would agree that everyone should know: "Columbus discovered America in 1492." Of what import is this "fact"? Most facts are oversimplifications of very complex events and when they are learned as facts, they lose all their interesting properties. What difference does it make when this event took place? What matters is that something took place, that we understand the events that led up to that something and the consequences that that something may have had for our lives. There may well be much controversy about Columbus's discovery of America, but there is less controversy over the fact that Columbus's act opened a chapter in the history of the world that had quite important ramifications. This would matter if you were thinking about Bosnia, Iraq, or the plight of the American Indians, for example, and that would be when you might want to learn it. Learn it at a different time, apart from any context, and it will be useless.

If there are no facts worth knowing, then what is worth knowing? First, there are skills, in particular basic skills such as reading, writing, and arithmetic. Also, there are other, less basic skills, such as speaking well, relating to others, understanding the world you live in. Processes, too, are worth knowing: political processes, psychological processes, physical processes, and economic processes. Understanding how things work so that you can work with them and make them work for you is also important. Cases are worth knowing. What is a case? For any subject matter in which a student shows an interest, it will be interesting for that student to hear stories that illustrate truths about that subject, exciting things about that subject, the experience of others with that subject, and so on. Getting computers to have a large "case base" in terms of which they can comprehend new cases and propose new generalizations turns out to be the critical issue in creating artificial intelligence.

One's own experience is, of course, worth knowing. Since we learn best by doing, students must be given real things to do from which they can acquire their own "case base." The best way to learn about a political process, for example, is to engage in one. So, a lot is worth knowing, but no facts are worth knowing in and of themselves. To become educated, you must direct your own education. To learn, you must do, and dwell on what prevents you from doing, so that you can change your knowledge base and try again. Seek out confusion in order to learn from it, and pay only as much attention to school as you have to, remembering that school and learning have little to do with each other.

ROGER C. SCHANK is a computer scientist and cognitive psychologist. At Northwestern University he is director of the Institute for the Learning Sciences, John Evans Professor of Electrical Engineering and Computer Science, and professor of psychology and of education and social policy. His primary work is attempting to understand how the mind works, focusing specifically on how language is understood, how human memory works, how learning occurs, how reasoning works, how to teach children, how to get computers to model people, and how to get computers to be useful to the man in the street.

He is author of fourteen books on creativity, learning, and artificial intelligence, including *The Creative Attitude: Learning to Ask and Answer the Right Questions,* with Peter Childers; *Dynamic Memory; Tell Me a Story;* and *The Connoisseur's Guide to the Mind.*

How Do We Communicate?

Dan Sperber

ommunicate. We humans do it all the time, and most of the time we do it as a matter of course, without thinking about it. We talk, we listen, we write, we read—as you are doing now— or we draw, we mimic, we nod, we point, we shrug, and, somehow, we manage to make our thoughts known to one another. Of course, there are times when we view communication as something difficult or even impossible to achieve. Yet, compared to other living kinds, we are amazingly good at it. Other species, if they communicate at all, have a narrow repertoire of signals that they use to convey again and again things like: "This is my territory," "Danger, run!" or "Ready for sex."

To communicate is to attempt to get someone to share your thoughts—well, at least some of them. But how can thoughts be shared? Thoughts aren't things out there in the open, to be sliced like cakes or used collectively like buses. They are strictly private affairs. Thoughts are born, live, and die inside our brains. They never truly come out of our heads (although we talk as if they did, but this is just a metaphor). The only thing that is ever produced by one person for another person to see or hear is behavior and the traces it leaves behind: movement, noise, broken twigs, ink spots, etc. These things aren't thoughts, they don't "contain" thoughts (that is just another metaphor), and yet some of these behaviors or traces serve to convey thoughts.

How is such communication possible? There is an old story—it dates back at least to the ancient Greek philosopher Aristotle—and no doubt you have heard it many times. What makes communication possible, so the story goes, is a common language. A language such as English is a kind of code in which sounds are associated to meanings and meanings to sounds. So, if Jill wants to communicate some meaning to Jack, she looks up in her mental grammar of English the sound associated to that particular meaning, and produces that sound for Jack to hear. Jack then looks up in *his* mental grammar the meaning associated with that particular sound. In that manner, Jack finds out what Jill had in mind. Of course, all this "looking up" is automatic and unconscious (except when you can't find your words, and become painfully aware of searching for them). Thanks to this double conversion—the encoding of meaning into sound, and the decoding of sound into meaning—Jill and Jack are now sharing a thought. Well, "sharing" may still be a metaphor, but at least we know now how to make good sense of it. Or do we?

The old "we communicate thanks to a common language" story is clever and simple. It would make a great explanation, if only it were true. Actually, some such story is true of most animal communication. Bees and monkeys have their own rudimentary codes, and whatever they communicate, they do so through encoding and decoding. Not so with us humans. True, we have our rich languages and many minor codes too, but—and this is where the old story breaks down—we manage to communicate much more than we encode and decode, and not just occasionally, but all the time. So, our having language is, at best, a mere part of the true story.

Let me illustrate. Imagine you are killing time at an airport. There is a woman standing nearby and you overhear her say to her companion, "It's late." You have heard and even uttered these very same words many times. Do you know what they mean? Of course. But do you know what the woman meant in uttering these words right now? Think about it. She might have been talking about a plane and meaning that it would arrive—or maybe depart—late. She may as well have been talking about a letter she was expecting, or about spring being late. She need not have been talking about anything in particular; she might just mean that it's late in the afternoon, or in the day, or in her life. Moreover, "late" is always relative to some schedule, or expectation; it might be late for lunch and yet early for supper. So she must have meant *late* relative to something, but what?

I could go on, but the point should be clear: Although you know perfectly well what the words the woman uttered mean, you don't know what *she* meant. Strangely enough, her companion does not seem puzzled. He seems to have understood her. And come to think of it, on the many occasions when you were the person told "It's late," you knew what the speaker meant. You didn't have to think about the many meanings that "It's late" might serve to convey. Is this sentence a special case? Not at all. Any English—or French, or Swahili—sentence may convey different meanings on different occasions, and might have served to illustrate the same point.

Because of such facts, linguists have found it necessary to distinguish "sentence meaning" from "speaker's meaning." Only linguists are interested in sentence meaning for its own sake. For the rest of us, sentence meaning is something we are generally unaware of. It is something we use unconsciously, as a mean toward our true end, which is to understand people, and to make ourselves understood. Speaker's meaning—the stuff we care about—always goes beyond sentence meaning: It is less ambiguous (although it may have ambiguities of its own); it is more precise in some ways, and often less precise in other ways; it has rich implicit content. Sentence meaning is but a sketch. We arrive at speaker's meaning by filling in this sketch.

How do we go from sentence meaning to speaker's meaning? How do we flesh out the sketch? In the past twenty years or so, it has become obvious that, in order to grasp a speaker's meaning, we make use of inference. *Inference* is just the psychologists' term for what we ordinarily call "reasoning." Like reasoning, it consists in starting from some initial assumptions and in arriving through a series of steps at some conclusion. Psychologists, however, are not just being pretentious in using a rarer word. When most of us talk of reasoning, we think of an occasional, conscious, difficult, and rather slow mental activity. What modern psychology has shown is that something like reasoning goes on all the time—unconsciously, painlessly, and fast. When psychologists talk of inference, they are referring first and foremost to this ever-present mental activity. Here, then, is how today's linguists and psychologists understand how one person understands what another person says. When you are told something, for instance, "It's late," first you *decode* the sentence meaning, and then you *infer* the speaker's meaning. All this, however, takes place so fast and so easily that it seems immediate and effortless.

How, then, should we revise our understanding of human communication? The first response is to stay as close as possible to the old coding-decoding theory. The updated story might go like this: What makes communication possible is the possession of a common language, as we always said; however, given human intelligence, you don't need to encode all your meaning, or to encode it exactly, in order to be understood. You can trust your audience to infer your full meaning from their knowledge of the situation, taken together with what you actually said. Why indeed say, "The plane on which your mother is coming is late, so late that we cannot wait for her any longer. I told you, we should have stayed at home," when saying, "It's late!" with the right tone of voice can convey all of this and more. The role of inference in communication is that of an optional add-on. All that is really needed for communication is a common language, but inference provides fast routines and shortcuts that are too effective to do without.

Many psychologists and linguists accept this updated version of the old story. Others don't. Trying to understand the kind of inference involved in communication has led some of us to turn the old story upside down. We now think that human communication is first and foremost a matter of inference and that language is the add-on. Here is the new story.

A million years ago, let's assume, our ancestors had no language at all. One of our ancestors, call him Jack, was watching an ancestress—call her Jill—picking berries. What did Jack understand of what Jill was doing? He might

have seen her behavior as a mere sequence of bodily movements, or he might have seen it as the carrying out of an
intention, perhaps the intention to gather berries for eating.
Understanding the behavior of an intelligent animal as the
carrying out of an intention is, in general, much more in-
sightful and useful than seeing it as mere movement. But
were our ancestors capable of recognizing intentions in one
another's behavior?

You have to be doubly intelligent to see the intelligence
in others. You need the ability to represent in your own mind
the mental representations of other creatures. You need, that
is, the ability to entertain representations of representations,
what, in our jargon, we call "meta-representations." Most
animals have no meta-representational capacity whatsoever.
In the world as they see it, there are no minds, only bodies.
Chimpanzees and other close relatives of ours seem to have
some rudimentary meta-representational capacity. As for
Jack, I bet he did perceive Jill's intention, and not just her
movements. In fact, he was probably gifted enough to infer
from her behavior not just her intention, but also one of her
beliefs: that those berries were edible.

If you are able to infer other people's beliefs from their
behavior, you can benefit from their knowledge and discover
facts of which you yourself have no direct experience. Jack
might not have known that these berries were edible, but
seeing Jill pick them gave him a good reason to believe that
they were. Even without the use of language or of communication, it may be possible to discover other people's
thoughts and to make them one's own.

Now, Jill was just as smart as Jack. She had noticed that
Jack was watching her, and she knew what he would infer
from her behavior. She may have liked Jack and felt glad

that her picking berries would serve two purposes instead of one: providing her with food, and providing Jack with information. In fact, it could be that Jill didn't really need the berries, and that her main purpose in picking them was to let Jack know that they were good to eat. Mind you, it could also be that she hated Jack, and, knowing that these particular berries were poisonous, she was trying to mislead him! We are coming closer to true communication with its tricks, but language is not yet in the picture. There is another big difference between Jill's attempt at informing or misinforming Jack and ordinary human communication. Ordinary communication is pursued openly. Here, on the other hand, Jack is not meant to realize that Jill is trying to alter his thoughts.

What if Jack understands that Jill's true intention in picking berries is to make him believe that they are edible? If he trusts Jill, he will believe her; if he doesn't, he won't. Now, what if Jill understands that Jack grasps her real purpose? Well, then, lo and behold, a world of possibilities opens! If Jack is capable of understanding that her purpose is to inform him, she might as well be open about it. Jill does not have to actually pick the berries anymore. All she must do is show Jack that she wants him to know that they are edible. She may, for that, resort to symbolic means.

Jill might, for instance, stare at the berries and then move her mouth, or she might mimic eating the berries. Jack would ask himself: Why does she do that? Once he recognized that she was doing that for his benefit, he wouldn't find it hard to infer her intention, or, in other words, her meaning. This is true overt communication, although still without language. All Jill does is give evidence of her intention, and all Jack does is infer what her intention is from the evidence she has given him. None of that evidence is linguistic or even codelike.

For creatures capable of communicating in this inferential manner, a language would be tremendously useful. Words are even better than mimicry for putting ideas in people's mind. If Jill had been able to utter just *eat,* or *good,* Jack could have inferred her intention, her full meaning, from her verbal behavior as easily as he did from her miming. With a richer language, Jill would have been able to give evidence of more complex meanings. Actually, in those days, our ancestors did not speak. However, their capacity for inferential communication created an environment in which language would come as a major advantage, and sure enough, a capacity for language evolved in the human species.

The new story, then, is that human communication is a by-product of human meta-representational capacities. The ability to perform sophisticated inferences about each other's states of mind evolved in our ancestors as a means of understanding and predicting each other's behavior. This, in turn, gave rise to the possibility of acting openly so as to reveal one's thoughts to others. As a consequence, the conditions were created for the evolution of language. Language made inferential communication immensely more effective. It did not change its character. All human communication, linguistic or nonlinguistic, is essentially inferential. Whether we give evidence of our thoughts by picking berries, by mimicry, by speaking, or by writing—as I have just done— we rely first and foremost on our audience's ability to infer our meaning.

DAN SPERBER is senior research scholar at the Centre National de la Recherche Scientifique and at the Ecole Polytechnique in Paris. Together with the British linguist Deirdre Wilson, he is author of *Relevance: Communication and Cognition,* in which they develop their groundbreaking and controversial approach to human communication; and *Relevance Theory,* which has inspired much novel research since. He is also author of *Rethinking Symbolism,* and *On Anthropological Knowledge.*

Minds, Brains, and Rosetta Stones

Steven Rose

~~~~~~~~~~ Our language is full of dichotomies: nature versus nurture;
genes versus environment; masculine versus feminine; hardware versus software; cognition versus affect; soul versus
body; mind versus brain. But do such divisions in our
thinking reflect real differences in the world outside us, or
are they the products of the intellectual history of our society—that is, are they ontological or epistemological?—and
note that this distinction, too, is dichotomous! One way of
answering that question is to ask whether other cultures and
societies make the same types of split. In the case of mind
versus brain, they clearly do not; according to the historian
of science Joseph Needham, Chinese philosophy and science,
for example, did not recognize any such distinction. And
while the mind/brain split is presaged in much of the Greco–
Judeo-Christian tradition, it was really given substantive
form only in the seventeenth century, at the birth of modern,
Western science. It was then that the Catholic philosopher

and mathematician René Descartes divided the universe into two sorts of stuff, the material and the mental. All the living, natural world around us was consigned, along with human-made technology, to the material; so, too, were human bodies. But to each human body was also attached a mind or soul, breathed into the body by God, and linked to it by way of an organ deep inside the brain, the pineal gland.

The split was useful in a number of ways. It justified human exploitation of other animals, for these were mere mechanisms and not worthy of more consideration than any other sort of machine; it exalted humanity's special place in the scheme of things, but only insofar as the human soul went; human bodies, too, could be exploited, and increasingly were through the American slave trade and as the industrial revolution of the eighteenth and nineteenth centuries gathered pace; souls could be left for Sunday pastoral care.

Cartesian dualism left its mark on medicine, especially medicine that dealt with the mind. Mental distress and disorder became dichotomized as either organic/neurological—something wrong with the brain—or functional/psychological—something wrong with the mind. These dichotomies persist today in much psychiatric thinking, so that therapy becomes divided between drugs to treat the brain and talk to treat the mind. The causes of these distresses are usually seen as lying either in the realm of the mind ("exogenous," as in a depression following on personal tragedy) or of the body ("endogenous," the product of faulty genes and disordered biochemistry).

But as the power and scale of modern science expanded since the seventeenth century, Descartes's uneasy compromise found itself increasingly under challenge. Newtonian physics ordered the motion of the planets and the falling of apples. Antoine-Laurent Lavoisier showed that human breathing was a process of chemical combustion, not different in principle from the burning of coal in a furnace. Nerves and muscles danced to the application of Luigi Galvani's electrical charges rather than the operation of some autonomous will. And Darwinian evolution placed humans on a par with other animals. A militant reductionism, a mechanical materialism, became the order of the day. In 1845, four rising French and German physiologists—Hermann Helmholtz, Carl Ludwig, Emil Du Bois-Reymond, and Ernst Brücke—swore a mutual oath to account for all bodily processes in physicochemical terms; in Holland, Jacob Moleschott went further, claiming that the brain secretes thought as the kidney secretes urine, that genius was a matter of phosphorus. For the English champion of Darwinism, Thomas Huxley, mind was to brain as the whistle was to the steam train.

More than a century later, such reductionism is the conventional wisdom of much science. Many believe the "most fundamental" of sciences is physics, followed by chemistry, biochemistry, and physiology; higher up the hierarchical scale are the "softer" sciences of psychology and sociology, and the goal of a unified science is seen as to collapse all these higher-order sciences into the "more fundamental" ones. Molecularly oriented scientists are openly contemptuous of the claims of the "softer" subjects. In 1975, E. O. Wilson opened his famous (or notorious, depending on your perspective) text *Sociobiology, the New Synthesis* with the claim that between them, evolutionary biology and neurobiology were about to make psychology, sociology, and economics irrelevant; ten years later, the doyen of molecular biology, Jim Watson, startled his audience at London's Institute of

Contemporary Arts with the claim that "in the last analysis there are only atoms. There's just one science, physics; everything else is social work."

So what of memories of childhood, pleasure in hearing a Beethoven string quartet, love, anger, the hallucinations of schizophrenia, a belief in God or a sense of injustice in the world—or even consciousness itself? Do we go with Descartes and consign them to the world of the mental and spiritual, themselves untouched by the carnal world though capable of touching it by twitching at the nerves? John Eccles, a Nobel prize-winner for his work on the physiology of synapses, the junctions between nerve cells, and like Descartes before him a committed Catholic and dualist, certainly believes this, for he has argued that there is a "liaison brain" in the left hemisphere through which the soul can tweak the synapses. Or do we align ourselves with Watson, Wilson, and their nineteenth-century forebears and go for the genes and dismiss the rest? As a colleague of mine in biochemistry put it at a conference for parents of "learning disabled" kids, is our task to show "how it is that disordered molecules lead to diseased minds?"

Well, let me set out my own position as a neuroscientist interested in the functioning of mind and brain. First, there is only one world, a material, ontological unity. The claim that there are two sorts of incommensurable stuff in the world, material and mental, leads to all sorts of paradoxes and is unsustainable. Without getting into long philosophical debates, the simple observations that manipulating brain biochemistry (for instance, by psychoactive drugs) alters mental perceptions, or that PET-scan imaging shows that specific regions of the brain use more oxygen and glucose when a person is lying quietly "mentally" solving a mathematical problem, show that while genius may be more complicated than a matter of phosphorus, what we describe as brain processes and as mental processes must be linked in some way. So monism, not dualism, rules.

But that does not put me into the Watson and Wilson camp. There is more to understanding the world than just enumerating the atoms within it. To start with, there are the organizing relations between the atoms. Consider a page of this book. You see it as a sequence of words, combining to make sentences and paragraphs. A reductionist analysis could decompose the words into their individual letters, the letters into the chemical constituents making up the black ink on white paper. Such an analysis would be comprehensive; it would tell you exactly what the composition of this page is; but it would not tell you about the meaning of the letters combined as words, sentences, and paragraphs. This meaning is only apparent at the higher level of analysis, a level at which the spatial distribution of the black ink on white paper, the pattern and nonrandom spatial order of the words as they appear on the page, and the sequential relationship of each sentence to the next come into view. To interpret the pattern, it is not chemistry but a knowledge of language that is required. The new higher level of organization thus needs its own science. For example, the study of liquids requires that we use such properties as cohesion and incompressibility to explain flow, vortex, and wave formation, none of which are properties of the molecules of which the liquids are made. Similarly, brains have properties such as memory storage and recall that are not found in single cells. These qualitatively different aspects of systems at different levels are emergent properties, and biology is full of them.

Furthermore, to make sense of the spatial order of the words on the page also demands a temporal order. In scripts derived from the Latin, one begins reading at the top left-hand corner of the page and continues to the bottom right. Reversing the order results in nonsense. Temporal, developmental order is a vital feature of higher-level organization and processes in a way that is not necessarily the case at lower levels and cannot be interpreted within a reductionist framework. Even more: The symbols on the page alone are insufficient; to understand the significance of a page of prose requires that we know something of the language and culture within which it has been composed and the purposes for which it has been written. (Is what's on this page, one can wonder, a fish taxonomy, a recipe for bouillabaisse, or an ode to the joys of Mediterranean culinary culture?) One very important principle of biological organization is indicated by this simple analogy. Nothing in biology makes sense except in the context of history; the history of the individual organism (that is, its development) and the history of the species of which that organism is a member (that is, evolution). Indeed, evolution can in some senses be regarded as the history of the emergent events that gave rise to the diversity of organisms of different forms and behaviors, which is such a distinctive feature of the living world.

Explaining the black squiggles on the white page of a book in terms of their chemistry helps in our understanding of their composition; however, it tells us nothing about their meaning as a set of ordered symbols on the page. Explaining is not the same as explaining away and no amount of chemical sophistication is going to eliminate the higher-order science that is needed to provide such meaning. Furthermore, and very practically, the naive reductionist program offered by Wilson and Watson simply doesn't work in practice. There are few simpler compound molecules than that of water—two hydrogen atoms and one oxygen atom combining to give H_2O. Yet all the resources of physics are not enough to enable one to predict the properties of this molecule from a knowledge of the quantum states of its constituent hydrogen and oxygen. Chemistry is never going to be collapsed into physics, even though physical principles and knowledge profoundly illuminate chemistry. And still less are sociology and psychology going to be collapsed into biochemistry and genetics.

So despite the ontological unity of the world, we are left, and always will be, with a profound epistemological diversity. In the well-worn analogy of the blind people and the elephant, there are many things to know and many ways of knowing. And we have many different languages that we use to describe what we know. Take a simple biological phenomenon, the twitch of a muscle in a frog's leg when an electric shock is applied to it or a motor nerve fires. For physiologists, this twitch can be described in terms of the structure and electrical properties of the muscle fibrils as seen in the microscope and read by a recording electrode placed on the muscle surface. For biochemists, muscle cells are largely composed of two types of protein, actin and myosin, long threadlike, interdigitating molecules; during the muscle twitch, the actin and myosin filaments slide across each other. In simple language, we are often tempted to say that the sliding filaments of actin and myosin "cause" the muscle twitch. But this is a loose and confusing use of language. The term *cause* implies that first one thing happens (the cause) and then another follows (the effect). But it is not the case that first the actin and myosin filaments slide across each other and then the muscle twitches. Rather, the sliding of the filaments is the same as the muscle twitch, but described in a different language. We can call it biochemese, to contrast it with physiologese.

Where does this leave the mind/brain dichotomy with which I began this discussion? Brain does not "cause" mind, as naive mechanical materialism would suggest (as the whistle to the steam train), nor are mind and brain two different sorts of thing, as Cartesian dualism argues. Rather, we have one thing, brain/mind, which we can talk about in two rather different languages, perhaps neurologese and psychologese.

An example: One of the most common types of mental distress in Europe and the United States today is depression. For many years, biologically oriented psychiatrists, social psychiatrists, and psychotherapists have been at loggerheads over the causes and treatments of depression. Is it, as the biological psychiatrists would claim, caused by disorders of neurotransmitter metabolism in the brain, or is it the result of intolerable pressures of day-to-day living? (Among the best predictors of depression is to be a low-income single mother living in inner-city housing in conditions of economic and personal insecurity.) If the former, then depression should be treated by drugs which affect neurotransmitter metabolism; if the latter, by alleviating the social and personal conditions which cause the distress or by enabling the person to cope better with them. This is the prescription offered by psychotherapy. But in my terms, these are not incompatible types of explanation or treatment. If biological psychiatry is right, depressed people have disordered neurotransmitters, and if psychotherapy works, then as one goes through psychotherapy and the depression lifts, the neurotransmitter disorder should correct itself. A couple of years ago, I set out to test this (I had to overcome quite a lot of hostility both from the biological psychiatrists and from the psychotherapists to do so) by measuring both psychiatric rating and the level of a particular neurotransmitter/enzyme system in the blood of patients at a London walk-in psychotherapy center. I followed the patients for a full year through their treatment. The results were quite clear-cut. Patients who entered the therapy feeling (and scoring on the rating scales) depressed had lower levels of the neurotransmitter than matched, apparently normal, control subjects. Within a few months of entering therapy, their depression rating improved and their neurotransmitter levels were back to normal values. The biochemical change and the talking therapy went hand-in-hand.

Mind language does not cause brain language, or vice-versa, anymore than a sentence in French causes a sentence in English, but they can be translated into each other. And just as there are rules for translating between French and English, so there are rules for translating between neurologese and psychologese. The problem for the mind/brain scientist then becomes one of deciphering these rules. How is this to be achieved?

Let me suggest an analogy. Pass through the massive neoclassical entrance to the British Museum in London, turn left through the shop, and pick your way through the throngs of tourists tramping the Egyptian and Assyrian galleries. A knot of people lean over a slab of black stone mounted at a slight angle to the floor. If you can interpose your body between the tourists and their miniature videocameras, you will see that the flat surface of the stone is divided into three sections, each covered with white marks. The marks in the top third are ancient Egyptian hieroglyphs; those in the middle are in a cursive script, demotic Egyptian; and if you had what used to be regarded as a "sound classical education" or have been to Greece on holiday, you will recognize the writing in the lower third as Greek. You are looking at the Rosetta stone, the text of a decree passed by a general council of Egyptian priests who assembled at Memphis on the Nile on the first anniversary of the coronation of King Ptolemy in 196 BCE. "Discovered" (in the sense that Europeans talk of artifacts of which they were previously unaware, irrespective of what the local population might have known of them) by a lieutenant of engineers in Napoleon's Egyptian expeditionary force in 1799, the stone became British booty of war with the French defeat, and was brought back to London and placed ritually among the great heap of spoils of ancient empires with which the British aggrandized themselves during their own century of imperial domination.

But the importance of the Rosetta stone lies not in its symbolism of the rise and fall of empires (even the Greek portion of its three scripts indicated that at the time it was carved, Egyptian power was in slow decline, and the rise of European preeminence was beginning). The fact that its three scripts each carry the same message, and that nineteenth-century scholars could read the Greek, meant that they could begin the task of deciphering the hitherto incomprehensible hieroglyphs that formed ancient Egypt's written language. The simultaneous translation offered by the Rosetta stone became a code-breaking device, and for me, it is a metaphor for the task of translation that we face in understanding the relationships between mind and brain.

Brain and mind languages have many dialects, but they are related in the same way as the scripts on the Rosetta stone. Where, however, can we find a code-breaking device? The psychotherapy example is suggestive, but in no way can it be rigorous enough to give us the code we need. Yet, it does provide one clue. A fundamental feature of experimental science is that it is easier to study change than stasis. If we can find a situation in which mind processes change, as when the depressed patients improve, and ask what is changing at the same time in brain language, then we can begin the process of mapping brain change on mind change. Until the development, over the last few years, of scanning techniques that provide windows into the functioning human brain—techniques like PET and MRI—looking inside the brain was only possible in experimental animals. For both humans and animals, one of the most clear-cut and simple examples of changing minds is provided when some new task or activity is learned and subsequently remembered. Such experiences of learning and remembering are much easier to study in the laboratory than the complexities of depression, and psychologists throughout this century, from Ivan Pavlov and B. F. Skinner on, have filled miles of library shelves with detailed training protocols to induce dogs to salivate to the sound of bells, rats to press levers to obtain food, and rabbits to blink their eyes to flashes of light. Now all we have to do is to show what happens inside the brain when such learning occurs. This is indeed what neuroscience labs around the world have been attempting over the past decade or so, and we are beginning to be able to tell quite a clear story in brain language of the mechanisms involved in learning; of how, during new experiences, new pathways are laid down in the brain inscribing the memory there much as traces are laid on a magnetic tape when music is recorded, and which can subsequently be played back just as the magnetic tape can.

Does this reduce the learning and memory to "nothing but" pathways in the brain? No more than the Beethoven quartet is reduced to the magnetic patterns on the tape. The sound of the music, its resonances as we hear and respond to it, are no more diminished or explained away by being recorded on tape or CD than is the validity and personal significance of our memories by being stored in new pathways in the brain. Rather, knowing the biology of how we learn and remember adds to our human appreciation of the richness of our own internal living processes. We need, and will always continue to need, both languages, of mind and brain, and the translation rules between them, to make sense of our lives.

STEVEN ROSE was educated at Cambridge (double first-class degree in biochemistry). His driving scientific interest in understanding the brain led him to take a Ph.D. at the Institute of Psychiatry in London. After periods of postdoctoral research at Oxford (Fell, New College), Rome (NIH fellow), and with the Research Council in London, in 1969 he became professor and chair of the department of biology at Britain's newly formed Open University, where, at age thirty-one, he was one of Britain's youngest full professors. At the Open University, he established and has directed ever since the Brain and Behavior Research Group and has focused his research on understanding the cellular and molecular mechanisms of learning and memory. His pioneering research in this area has led to the publication of some two hundred and fifty research papers and receipt of various international honors and medal awards.

As well as his research papers, Steven Rose has written or edited fourteen books, including *No Fire, No Thunder* (with Sean Murphy and Alastair Hay); *Not in Our Genes* (with Richard Lewontin and Leo Kamin); *Molecules and Minds;* and most recently, *The Making of Memory,* which won the 1993 Rhone-Poulenc Science Book Prize.

Rose's concerns have always included not merely the doing and communicating of science, but also the wider social and ideological framework of science and its social uses. These concerns led him to play a central role in the 1960s and 1970s, with the feminist sociologist Hilary Rose, in establishing the British Society for Social Responsibility in Science. He has also recently taken a strong public stand on the issue of animal rights and animal experimentation.

Study Talmud

David Gelernter

There are lots of unscientific texts in the world, but you'd be hard-pressed to find a less scientific one than the Talmud. Still, I am quite serious about my recommendation. Let me explain.

To study Talmud is to study reading. The Talmud is the ultimate challenge to close readers. Whether you do the thing right and study for a decade or merely skim the surface for a year or two, you cannot emerge from Talmud study without a transformed understanding of how to squeeze, distill, coax, urge, cajole, boil, or otherwise extract meaning from a difficult text. Scientists nowadays rarely know how to read seriously. They are accustomed to strip-mining a paper to get the facts out and then moving on, not to mollycoddling the thing in search of nuances; there probably aren't any. Ordinarily, reading is irrelevant to science. But there are exceptions, and they can be important.

They are especially important to computer science, because computer science is a field that isn't quite sure what it is all about. It suspects that it is important, and onto something big. But it is still puzzling over its mandate. Computer science is the science of software, and software is the strangest stuff on earth. We still don't understand it very well, and we don't yet understand the boundaries and methods of software science.

Close reading is important to computer science, because the authors of some of its landmark papers were still thrashing over, in their own minds, questions about the fundamental character of the topic. The papers say one thing on the surface; between the lines, they address basic questions that are still unresolved.

Among the most basic of those questions is the relationship between software and mathematics. From the invention of the first widely used programming language in 1957, many computer scientists have suspected that software is a lot like mathematics or that it *is* mathematics. Some computer scientists believe that writing a program and constructing a mathematical proof are equivalent and interchangeable activities, two equally fine ways to kill a Sunday afternoon. They furthermore believe that the tools we use to write software—programming languages, mainly—ought to be *defined* mathematically, using a series of equations.

Others insist that these claims are wrong. (As, in fact, they are.) They argue that software is not in the least like mathematics. A piece of mathematics is a form of communication among people. A piece of software is a blueprint for a special kind of machine, a "virtual" or "embodied" machine. When you run the program on a computer, your computer temporarily *becomes* the machine for which your program serves as the blueprint. Designing a program is nothing like doing a proof; it is much more like designing a car. And the tools we use to write software ought to be defined (this school furthermore holds) in clear, simple English, not mathematically in a series of equations.

This dispute gets to the heart of what software and computer science are all about. It is also one battle in a war that rages up and down modern intellectual history, the war between the mathematical and the physical worldviews. Battles have been fought in civil engineering: Do we design bridges based on mathematical models, or based on experience, aesthetics, and intuition? They have been fought in physics— is mathematics, or seat-of-the-pants physical intuition, the ultimate guiding light? They have been fought in economics and in other social sciences; and they are basic to computer science.

Now, let's consider the value of careful reading. It does not (in the cases I discuss) reward us with decisive opposition-destroying ammunition; merely with subtle insights that are worth having.

Algol 60 is the most important programming language ever invented, and the revised Algol 60 report of 1963 is for several reasons arguably computer science's most important paper. Algol 60 itself is decisively important; furthermore, the report is widely believed to be a masterpiece of lucid writing. It uses clear, simple English to explain what Algol 60 means and what it does. If you are going to use English and not equations to explain a programming language, it's unlikely that you will be able to do a better job than this paper did.

And that is precisely why it is so important to the "mathematics" school that the Algol 60 report *is in some ways ambiguous and incomplete. "That does it!"* they say. In this classic paper, English took its very best shot, and fell short. Mathematics is the way to go: QED.

One champion of the mathematical approach puts it this

way: "Both users and implementors of programming languages need a description that is comprehensible, unambiguous and complete. . . . Almost all languages have been defined in English. Although these descriptions are frequently masterpieces of apparent clarity they have nevertheless usually suffered from both inconsistency and incompleteness."[1] The author goes on to mention the famous Algol 60 report,[2] and to cite a well-known paper whose name is sufficiently incriminating to seal the case all by itself: "The Remaining Trouble Spots in Algol 60."[3]

One thing you learn in Talmud study is always look up the sources. When we examine the "Trouble Spots" paper, we find interesting reflections—such as, "The following list [of trouble spots] is actually more remarkable for its shortness than its length." Or, in conclusion, "The author has tried to indicate every known blemish in [the Algol 60 report]; and he hopes that nobody will ever scrutinize any of his own writings as meticulously."

The more we mull it over, the clearer it becomes that this "Trouble Spots" paper is evidence of the superb *clarity,* not the insufficiency, of the Algol 60 report and its English-language definitions. The report introduced a complex, unprecedented, and wholly original tool. That it succeeded in doing so with as few as the "Trouble Spots" paper documents is indeed (as the author asserts) remarkable. Mulling in this vein, we turn back to the report itself and notice its epigraph, concluding, *Wovon man nicht reden kann, darüber muss man schweigen.* One of the century's most famous philosophical pronouncements, the close of Ludwig Wittgenstein's *Tractatus:* "On matters about which one cannot speak, one must remain silent." Now the report *might* have been attempting to say: "We realize there are gaps in what follows, but the English language let us down. It was just not up to the task of closing them." Such a sentiment would have been vaguely related to Wittgenstein's actual contention, which had to do with the limits of any language's expressivity. But a far more likely interpretation is: "We realize there are gaps in this paper; but when we had nothing intelligent to say, we shut up."

The report's authors, on this reading, were under no delusions that their work was perfect. They produced no "masterpiece of *apparent* clarity" (my italics) that turned out to be a cheesy second-rate job as soon as the customers took a careful look. The shortfalls in the report—such as they are— are due, on this reading, not to the English language, but to the authors.

Are you quite sure, having made this small attempt at careful reading, that you want to conclude that English has flunked the test and mathematics is our only remaining choice?

My other example has to do, again, with the nature of software and the war over mathematics. It introduces another issue as well. Computer science is a fad-ridden, fad-driven field. In being so, it merely reflects (once again) the nature of modern intellectual life. My goal isn't to condemn computer science or urge it to change its ways. It is merely to face facts. We are a terribly unreflective field. Cattle are born to stampede and so are we, but humans have the advantage of being able to mull things over, along the lines of—"Well, here goes another stampede. Pretty much like the last one, no doubt." Presumably, we gain a bit of perspective if we trouble to do so.

Niklaus Wirth published the first description of his language Pascal in 1971, and within a few years, it was a dominant factor on the computer science scene. Pascal today is no longer used as widely as it once was for software development, but it remains the world's most popular teaching language, and a first-rate historical milestone. And Wirth is an authentic hero of the field.

In 1975, Wirth published an assessment of the highly successful Pascal Project.[4] To close readers, the paper is intriguing not so much for what it says about science and engineering as for what it says (between the lines) about how science and engineering are done. Superficially, the paper is a sober appraisal of the project, but between the lines, we discover the transcript of a bizarre show trial in which Wirth defends himself against a series of brand-new crimes.

Programs are complicated structures—in extreme cases, arguably the most complicated known to man—and it's very hard to get them to work just right. Elaborate, painstaking testing of the finished product is required in order to convince people that a program actually works. In the early 1970s, a new school of thought arose. It held that, since a program is mathematics, it ought to be possible to prove mathematically that a program worked. You wouldn't have to run the program; you would merely look it over carefully and construct a proof. There would be no nonsense about testing. Your software would be guaranteed perfect out of the box. This approach was called "program verification."

It sounds like a great idea, but it is unworkable in practice. It's as if you had designed a new car and were determined to prove just on the basis of the engineering drawings that everything was correct and would work perfectly. You'd be able to sell the first model off the assembly line to an eager customer without bothering to test-drive a prototype even once. That would be nice if you could make it work, but you can't, and software doesn't work that way either.

When Wirth designed Pascal, program verification was just gathering steam. When he looked back several years later, it was a full-blown monster fad. Wirth's retrospective complaints about his handiwork don't deal, for the most part, with people's actual *experiences* using Pascal; they mostly seem to turn on program verification. It's hard not to be reminded of a newly minted Marxist rising to confess his sins and repent of his past life. The details are technical, but I will mention a few briefly. Pascal streamlined and simplified Algol's operator precedence rules, but that turned out to be a mistake—not because programmers complained, but because of "the growing significance" of "program verification." Pascal included something called "the goto statement," which was unfortunate—not because programmers didn't like it, but because of problems associated with "the verifiability of programs." Programmers were very pleased with Pascal's flexible type system, but the really important thing about this innovation has to do not with their enthusiasm but with "invariants needed for the verification of programs."

I mentioned that Wirth is a hero. That is true not only because of the importance of his work in general, but because of this very paper. The "free-type union" was one aspect of Pascal that gave program verifiers fits. They demanded repeatedly that it be dragged out of the language and shot. Wirth acknowledges that the free-type union poses grave dangers, but he refuses to surrender it to the mob. "There seem to exist occasions," he says circumspectly, where the alternative to the insidious free-type union "is insufficiently flexible." The free-type union stays. Its comrades rejoice and the mob gnashes its teeth, but toward sunrise is seen to disperse by a relieved populace.

We ought, as a field, to know ourselves better than we do. Careful reading of this paper teaches us something important about the ebb and flow of fashion in computer science. Pascal is still around, but program verification is (more or less) dead. At least in the form that the 1970s cared about.

A word about the Talmud itself. It is arranged as a base text with a deep stack of commentaries. Each commentary in the stack is expected to refer to (or at least to be aware of) not only the base text, but all other commentaries before it. The base text itself is a two-layer structure: a terse "basic" base called the Mishnah, and a lengthy, loosely organized commentary called the Gemara. To study the Talmud is to study close reading at the highest possible level of analytic precision and rigor. A term or two of Talmud study would stand any future scientist in good stead.

Sociological postscript: An innocent observer might say, Wonderful! "Multiculturalism" is a big deal on our nation's campuses and in the public schools. Exposing a broader slice of students to the beauty and depth of Jewish culture must be exactly what multiculturalism is all about, right?

Don't hold your breath. If multiculturalism were a serious scholarly movement and not the tendentious clown act it turns out to be in reality, Jewish culture might be one of its greatest discoveries. As is, multiculturalism is merely the latest embodiment of the down-home "Roll Over Beethoven!" anti-intellectualism that has always been part of American life. The only difference in this brand of anti-intellectualism is that university presidents are its biggest fans. History is full of surprises.

References

1. M.J.C. Gordon, *The Denotational Description of Programming Languages.* New York: (Springer-Verlag, 1979).

2. P. Nauer, ed., "Report on the Algorithmic Language Algol 60." *Communications of the ACM.* Vol. 6, 1 (1963): pp. 1–17.

3. D. E. Knuth, "The Remaining Trouble Spots in Algol 60." *Communications of the ACM.* Vol. 10, 10 (1967): pp. 611–17.

4. N. Wirth, "An Assessment of the Programming Language Pascal." *IEEE Transactions on Software Engineering.* (1975): pp. 192–98.

DAVID GELERNTER, associate professor of computer science at Yale University, is a leading figure in the third generation of artificial-intelligence scientists. As a graduate student thirteen years ago, Gelernter wrote a landmark programming language called Linda that made it possible to link computers together to work on a single problem. He has since emerged as one of the seminal thinkers in the field known as parallel, or distributed, computing. He is author of *Mirror Worlds, The Muse in the Machine,* and *1939: The Lost World of the Fair.*

Identity in the Internet

Sherry Turkle

The technologies of our everyday lives change the way we see the world. The painting and photograph appropriated nature. When we look at sunflowers or waterlilies, we see them through the prism of how they have been painted. When we marry, the ceremony and the ensuing celebration "produce" photographs and videotapes that displace the event, and become our memories of it. Computers, too, cause us to construct things in new ways. With computers, we can simulate nature in a program or leave nature aside and build "second natures" limited only by our powers of imagination and abstraction. At present, one of the most dramatic of these "second natures" is found on interactive computer environments known as MUDs.

In MUDs (short for multi-user dungeons or multi-user domains), players logged in from all over the world, each at his or her individual machine, join on-line communities,

communities that exist only in the computer. MUDs are social virtual realities in which hundreds of thousands of people participate. They communicate with one another individually and in groups. Access is neither difficult nor expensive. One only needs internet access, now commercially available.

You join a MUD through a "telnet" command that links your networked computer with another. When you start, you create a character or several characters; you specify their genders and other physical and psychological attributes. Other players in the MUD can see this description. It becomes your character's self-presentation. The created characters need not be human, and there are more than two genders. Indeed, characters may be multiple ("a swarm of bees") or mechanical (you can write and deploy a program in the game that will present itself as a person or, if you wish, as a robot). Some MUDs have themes; some MUDs are free-form. At the time of this writing in 1994, there are over three hundred available MUDs on the internet.

On some of these MUDs, players are invited to help build the computer world itself. Using relatively simple programming languages, they can make a "room" in the game space where they are able to set the stage and define the rules. That is, they make objects in the "second nature" and specify how they work. An eleven-year-old player builds a room she calls "the condo." It is beautifully furnished; she has created magical jewelry and makeup for her dressing table. When she visits the condo, she invites her friends, she chats, orders pizza, and flirts. Other players have more varied social lives: They create characters who have casual and romantic sex, hold jobs, attend rituals and celebrations, fall in love, and get married. To say the least, such goings-on are gripping: "This is more real than my real life," says a character who turns out to be a man playing a woman who is pretending to be a man.

Since MUDs are authored by their players, they are a new form of collaboratively written literature which has much in common with performance art, street theater, improvisational theater, commedia del'arte, and script writing. But MUDs are something else, as well. As players participate in MUDs, they become authors not only of text, but of themselves, constructing selves through social interaction.

On MUDs, the obese can be slender, the beautiful can be plain. The anonymity of MUDs (you are known only by the name you gave your characters) provides ample room for individuals to express unexplored "aspects of the self." The games provide unparalleled opportunities to play with one's identity and to "try out" new ones. This aspect of their emotional power is well captured by the player who said:

> On a MUD, you can be whoever you want to be. You can completely redefine yourself if you want. You can be the opposite sex. You can be more talkative. You can be less talkative. Whatever. You can just be whoever you want really, whoever you have the capacity to be. You don't have to worry about the slots other people put you in as much. It's easier to change the way people perceive you, because all they've got is what you show them. They don't look at your body and make assumptions. They don't hear your accent and make assumptions. All they see is your words. And it's *always* there. Twenty-four hours a day you can walk down to the street corner and there's gonna be a few people there who are interesting to talk to, if you've found the right MUD for you.

In traditional role-playing games in which one's physical body is present, one steps in and out of a character; MUDs, in contrast, offer a parallel life. The boundaries of the game are fuzzy; the routine of playing them becomes part of their players' real lives. MUDs blur the boundaries between self and game, self and role, self and simulation. One player says, "You are what you pretend to be . . . you are what you play." But people don't just become who they play; they play who they are or who they want to be. Twenty-six-year-old Diane says, "I'm not one thing, I'm many things. Each part gets to be more fully expressed in MUDs than in the real world. So even though I play more than one self on MUDs, I feel more like 'myself' when I'm MUDding." Players sometimes talk about their real selves as a composite of their characters, and sometimes talk about their MUD characters as means for working on their real-world (RL) lives.

Many MUD players work with computers all day at their "regular" jobs. As they play on MUDs, they will periodically put their characters to "sleep," remain logged on to the game, but pursue other activities. From time to time, they return to the game space. In this way, they break up their workdays and experience their lives as a "cycling through" between the RL and a series of simulated ones.

This kind of interaction with MUDs is made possible by the existence of what have come to be called "windows" in modern computing environments. Windows are a way of working with a computer that makes it possible for the machine to place you in several contexts at the same time. As a user, you are attentive to only one of the windows on your screen at any given moment, but in a certain sense, you are a presence in all of them at all times.

Doug is a Dartmouth College junior majoring in business for whom a MUD represents one window and RL represents another. Doug says, "RL is just one more window and it not necessarily my best window."

Doug plays four characters distributed across three different MUDs. One is a seductive woman. One is a macho, cowboy type whose self-description stresses that he is a "Marlboros rolled in the T-shirt sleeve" kind of guy. Then there is a rabbit of unspecified gender who wanders its MUD introducing people to each other, a character he calls "Carrot." Doug says, "Carrot is so low key that people let it be around while they are having private conversations. So I think of Carrot as my passive, voyeuristic character." Doug tells me that this "Carrot" has sometimes been mistaken for a "bot"—a computer program on the MUD because its passive, facilitating presence strikes many as the kind of personae of which a robot would be capable.

Doug's third and final character is one whom he plays only on a furry MUD (these are MUDs known as places of sexual experimentation where all the characters are furry animals). "I'd rather not even talk about that character, because its anonymity there is very important to me," Doug says. "Let's just say that on furry MUDs, I feel like a sexual tourist." Doug talks about playing his characters in "windows," and says that using windows has enhanced his ability to "turn pieces of my mind on and off."

> I split my mind. I'm getting better at it. I can see myself as being two or three or more. And I just turn on one part of my mind and then another when I go from window to window. RL seems like one more window, and it is not usually my best window.

The development of the windows metaphor for computer interfaces was a technical innovation motivated by the desire to get people working more efficiently by "cycling through" different applications much as time-sharing computers cycle through the computing needs of different people. But in practice, windows have become a potent metaphor for thinking about the self as a multiple, distributed, "time-sharing" system. The self is no longer simply playing different roles in different settings, something that people experience when, for example, one wakes up as a lover, makes breakfast as a mother, and drives to work as a lawyer. The life practice of windows is of a distributed self that exists in many worlds and plays many roles at the same time. MUDs extend the metaphor—now RL itself, as Doug said, can be just "one more window."

A comment from a college student about MUDding illustrated that the MUD "window," however, offers special possibilities: "I can talk about anything on the MUD," he said. "The computer is sort of practice to get into closer relationships with people in real life." MUDs provide him with what the psychoanalyst Erik Erikson called a "psychosocial moratorium," which Erikson saw as part of a healthy adolescence. Although the term *moratorium* implies a "time-out," what Erikson had in mind was not withdrawal. On the contrary, he saw the moratorium as a time of intense interaction with people and ideas, a time of passionate friendships and experimentation. The moratorium is not on significant experiences but on their *consequences*. It is a time during which one's actions are not "counted." Freed from consequence, experimentation becomes the norm rather than a brave departure. Consequence-free experimentation facilitates the development of a personal sense of what gives life meaning that Erikson called "identity."

Erikson developed these ideas about the importance of a moratorium during the late 1950s and early 1960s. At that time, the notion corresponded to a common understanding of what "the college years" were about. Today, thirty years later, the idea of the college years as a consequence-free "time-out" seems of another era. College is preprofessional, and AIDS has made consequence-free sexual experimentation an impossibility. The years associated with adolescence no longer seem a "time-out." But if our culture no longer offers an adolescent moratorium, virtual communities do. It is part of what makes them seem so attractive.

Erikson's ideas about adolescence as a time of moratorium in the service of developing identity were part of a larger theory of stages of the life cycle. The stages were not to suggest rigid sequences, but descriptions of what people ideally need to achieve before they can easily move ahead to another developmental task. So, for example, identity development in adolescence would ideally precede the development of intimacy in young adulthood. In real life, however, people usually have to move on with incompletely resolved "stages." They do the best they can, making do with whatever materials they have at hand to get as much as they can of what they have "missed." MUDs are dramatic examples of how technology can play a role in these dramas of self-reparation. Time in cyberspace reworks the notion of moratoria, because it now exists in an always-available "window."

In assaulting our traditional notions of identity and authenticity, cyberspace challenges the way we think about responsibility as well. Wives and husbands need to decide whether a spouse is unfaithful if he or she collaborates on sexually explicit scenes with an anonymous character in cyberspace. Should it make a difference to a wife if unbeknownst to her husband, his cyberspace "girlfriend" turns out to be a nineteen-year-old male college freshman? What if she is an infirm eighty-year-old man in a nursing home? More disturbingly, if she is a twelve-year-old girl? Or a twelve-year-old boy? Disturbing in another sense, what if she turns out to be a skillfully written computer program, a "sex talk expert system," trained to participate in romantic encounters in cyberspace? For in MUDs, while some people are choosing to play the roles of machines, computer programs present themselves as people. MUDs blur the line between games and life and cause many other of our well-calibrated compasses to spin. People use concrete materials to think through their concerns, small and large, personal and social. MUDs are objects to think with for thinking about the self in a culture of simulation. Their citizens are our pioneers.

SHERRY TURKLE, professor of sociology of sciences at Massachusetts Institute of Technology, is a graduate and affiliate member of the Boston Psychoanalytic Society and a licensed clinical psychologist. She has written numerous articles on psychoanalysis and culture and on the "subjective side" of people's relationships with technology, especially computers. Her work on computers and people has been widely written about in both the academic and popular press, including *Time, Newseek,* and *US News and World Report.* She has spoken about the psychological and cultural impact of the computer as a guest on numerous radio and television shows, including *Nightline, The Today Show,* and *20/20.* She is the author of *Psychoanalytic Politics: Jacques Lacan and Freud's French Revolution; The Second Self: Computers and the Human Spirit;* and the forthcoming *Life on the Screen: Identity in the Age of the Internet.*

Part Five

Cosmos

What Is Time?

Lee Smolin

E very schoolchild knows what time is. But, for every schoolchild, there is a moment when they first encounter the paradoxes that lie just behind our everyday understanding of time. I recall when I was a child being struck all of a sudden by the question of whether time could end or whether it must go on forever. It must end, for how can we conceive of the infinity of existence stretching out before us if time is limitless? But if it ends, what happens afterward?

I have been studying the question of what time is for much of my adult life. But I must admit at the beginning that I am no closer to an answer now than I was then. Indeed, even after all this study, I do not think we can answer even the simple question: "What sort of thing is time?" Perhaps the best thing I can say about time is to explain how the mystery has deepened for me as I have tried to confront it.

Here is another paradox about time which I began to worry about only after growing up. We all know that clocks measure time. But clocks are complex physical systems and hence are subject to imperfection, breakage, and disruptions of electrical power. If I take any two real clocks, synchronize them, and let them run, after some time, they will always disagree about what time it is.

So which of them measures the real time? Indeed, is there a single, absolute time which, although measured imperfectly by any actual clock, is the true time of the world? It seems there must be, otherwise, what do we mean when we say that some particular clock runs slow or fast? On the other hand, what could it mean to say that something like an absolute time exists if it can never be precisely measured?

A belief in an absolute time raises other paradoxes. Would time flow if there were nothing in the universe? If everything stopped, if nothing happened, would time continue?

On the other hand, perhaps there is no single absolute time. In that case, time is only what clocks measure and, as there are many clocks and they all, in the end, disagree, there are many times. Without an absolute time, we can only say that time is defined relative to whichever clock we choose to use.

This seems to be an attractive point of view, because it does not lead us to believe in some absolute flow of time we can't observe. But it leads to a problem, as soon as we know a little science.

One of the things physics describes is motion, and we cannot conceive of motion without time. Thus, the notion of time is basic for physics. Let me take the simplest law of motion, which was invented by Galileo and Descartes, and formalized by Isaac Newton: A body with no forces acting on it moves in a straight line at a constant speed. Let's not worry here about what a straight line is (this is the problem of space, which is perfectly analogous to the problem of time, but which I won't discuss here). To understand what this law is asserting, we need to know what it means to move at a constant speed. This concept involves a notion of time, as one moves at a constant speed when equal distances are covered in equal times.

We may then ask: With respect to which time is the motion to be constant? Is it the time of some particular clock? If so, how do we know which clock? We must certainly choose because, as we observed a moment ago, all real clocks will eventually go out of synchronization with one another. Or is it rather that the law refers to an ideal, absolute time?

Suppose we take the point of view that the law refers to a single, absolute time. This solves the problem of choosing which clock to use, but it raises another problem, for no real, physical clock perfectly measures this imagined, ideal time. How could we truly be sure whether the statement of the law is true, if we have no access to this absolute, ideal time? How do we tell whether some apparent speeding up or slowing down of some body in a particular experiment is due to the failure of the law, or only to the imperfection of the clock we are using?

Newton, when he formulated his laws of motion, chose to solve the problem of which clock by positing the existence of an absolute time. Doing this, he went against the judgments of his contemporaries, such as Descartes and Gottfried Leibniz, who held that time must be only an aspect of the relationships among real things and real processes in the world. Perhaps theirs is the better philosophy, but as Newton knew better than anyone at the time, it was only if one believed in an absolute time that his laws of motion, including the one we have been discussing, make sense. Indeed, Albert Einstein, who overthrew Newton's theory of time, praised Newton's "courage and judgment" to go against what is clearly the better philosophical argument, and make whatever assumptions he had to make to invent a physics that made sense.

This debate, between time as absolute and preexisting and time as an aspect of the relations of things, can be illustrated in the following way. Imagine that the universe is a stage on which a string quartet or a jazz group is about to perform. The stage and the hall are now empty, but we hear a ticking, as someone has forgotten, after the last rehearsal, to turn off a metronome sitting in a corner of the orchestra pit. The metronome ticking in the empty hall is Newton's imagined absolute time, which proceeds eternally at a fixed rate, prior to and independently of anything actually existing or happening in the universe. The musicians enter, the universe all of a sudden is not empty but is in motion, and they begin to weave their rhythmic art. Now, the time that emerges in their music is not the absolute preexisting time of the metronome; it is a relational time based on the developing real relationships among the musical thoughts and phrases. We know this is so, for the musicians do not listen to the metronome, they listen to one another, and through their musical interchange, they make a time that is unique to their place and moment in the universe.

But, all the while, in its corner the metronome ticks on, unheard by the music makers. For Newton, the time of the musicians, the relational time, is a shadow of the true, absolute time of the metronome. Any heard rhythm, as well as the ticking of any real physical clock, only traces imperfectly the true absolute time. On the other hand, for Leibniz and other philosophers, the metronome is a fantasy that blinds us to what is really happening; the only time is the rhythm the musicians weave together.

The debate between absolute and relational time echoes

down the history of physics and philosophy, and confronts us now, at the end of the twentieth century, as we try to understand what notion of space and time is to replace Newton's.

If there is no absolute time, then Newton's laws of motion don't make sense. What must replace them has to be a different kind of law that can make sense if one measures time by any clock. That is, what is required is a democratic rather than an autocratic law, in which any clock's time, imperfect as it may be, is as good as any other's. Now, Leibniz was never able to invent such a law. But Einstein did, and it is indeed one of the great achievements of his theory of general relativity that a way was found to express the laws of motion so that they make sense whichever clock one uses to embody them with meaning. Paradoxically, this is done by eliminating any reference to time from the basic equations of the theory. The result is that time cannot be spoken about generally or abstractly; we can only describe how the universe changes in time if we first tell the theory exactly which real physical processes are to be used as clocks to measure the passage of time.

So, this much being clear, why then do I say that I do not know what time is? The problem is that general relativity is only half of the revolution of twentieth-century physics, for there is also the quantum theory. And quantum theory, which was originally developed to explain the properties of atoms and molecules, took over completely Newton's notion of an absolute ideal time.

So, in theoretical physics, we have at present not one theory of nature but two theories: relativity and quantum mechanics, and they are based on two different notions of time. The key problem of theoretical physics at the present moment is to combine general relativity and quantum mechanics into one single theory of nature that can finally replace the Newtonian theory overthrown at the beginning of this century. And, indeed, the key obstacle to doing this is that the two theories describe the world in terms of different notions of time.

Unless one wants to go backward and base this unification on the old, Newtonian notion of time, it is clear that the problem is to bring the Leibnizian, relational notion of time into the quantum theory. This is, unfortunately, not so easy. The problem is that quantum mechanics allows many different, and apparently contradictory, situations to exist simultaneously, as long as they exist in a kind of shadow or potential reality. (To explain this, I would have to write another essay at least as long as this one about the quantum theory.) This applies to clocks as well, in the same way that a cat in quantum theory can exist in a state that is at the same time potentially living and potentially dead, a clock can exist in a state in which it is simultaneously running the usual way and running backward. So, if there were a quantum theory of time, it would have to deal not only with freedom to choose different physical clocks to measure time, but with the simultaneous existence, at least potentially, of many different clocks. The first, we have learned from Einstein how to do; the second has, so far, been too much for our imaginations.

So the problem of what time is remains unsolved. But it is worse than this, because relativity theory seems to require other changes in the notion of time. One of them concerns my opening question, whether time can begin or end, or whether it flows eternally. For relativity is a theory in which time can truly start and stop.

One circumstance in which this happens is inside of a

black hole. A black hole is the result of the collapse of a massive star, when it has burned all its nuclear fuel and thus ceased to burn as a star. Once it is no longer generating heat, nothing can halt the collapse of a sufficiently massive star under the force of its own gravity. This process feeds on itself, because the smaller the star becomes, the stronger the force by which its parts are mutually attracted to one another. One consequence of this is that a point is reached at which something would have to go faster than light to escape from the surface of the star. Since nothing can travel faster than light, nothing can leave. This is why we call it a black hole, for not even light can escape from it.

However, let us think not of this, but of what happens to the star itself. Once it becomes invisible to us, it takes only a short time for the whole star to be compressed to the point at which it has an infinite density of matter and an infinite gravitational field. The problem is, what happens then? The problem, indeed, is what, in such a circumstance, "then" might mean. If time is only given meaning by the motion of physical clocks, then we must say that time stops inside of each black hole. Because once the star reaches the state of infinite density and infinite gravitational field, no further change can take place, and no physical process can go on that would give meaning to time. Thus, the theory simply asserts that time stops.

The problem is in fact even worst than this, because general relativity allows for the whole universe to collapse like a black hole, in which case, time stops everywhere. It can also allow for time to begin. This is the way we understand the big bang, the most popular theory, currently, of the origin of the universe.

Perhaps the central problem that those of us who are trying to combine relativity and quantum mechanics think about is what is really happening inside a black hole. If time really stops there, then we must contemplate that all time, everywhere, comes to a stop in the collapse of the universe. On the other hand, if it does not stop, then we must contemplate a whole, limitless world inside each black hole, forever beyond our vision. Moreover, this is not just a theoretical problem, because a black hole is formed each time a massive enough star comes to the end of its life and collapses, and this mystery is occurring, somewhere in the vast universe we can see, perhaps one hundred times a second.

So, what is time? Is it the greatest mystery? No, the greatest mystery must be that each of us is here, for some brief time, and that part of the participation that the universe allows us in its larger existence is to ask such questions. And to pass on, from schoolchild to schoolchild, the joy of wondering, of asking, and of telling each other what we know and what we don't know.

LEE SMOLIN, a theoretical physicist, is professor of physics and member of the Center for Gravitational Physics and Geometry at Pennsylvania State University. Together with Abhay Ashtekar and Roger Penrose, he holds a five-year National Science Foundation grant which supports their work in quantum gravity. In addition to being considered one of the premier scientists working in the field of quantum gravity, he has also made contributions to cosmology, particle physics, and the foundations of quantum mechanics.

Smolin is perhaps best known for a new approach to the quantization of general relativity, and as such, he has been identified as a leader of one of the most promising new directions currently being pursued in science. He has also been working on a proposal for applying evolutionary theory to cosmology, which has received a great deal of press, including articles in *The Independent, New Scientist,* and *Physics World* as well as two programs on BBC World Service. He is the author of more than fifty scientific papers and several articles for general audiences, and is at work on a popular book about quantum gravity.

Learning What Is, from What Doesn't

Alan H. Guth

W hen Alice protested in *Through the Looking Glass* that "one *can't* believe impossible things," the White Queen tried to set the issue straight. "I daresay you haven't had much practice," she said. "When I was your age, I always did it for half an hour a day. Why sometimes I've believed as many as six impossible things before breakfast."

Although science is in principle the study of what is possible, the advice of the White Queen is on target. No one yet understands the laws of nature at their most fundamental level, but the search for these laws has been both fascinating and fruitful. And the view of reality that is emerging from modern physics is thoroughly reminiscent of Lewis Carroll. While the ideas of physics are both logical and extremely beautiful to the people who study them, they are completely at odds with what most of us regard as "common sense."

Of all the "impossibilities" known to science, probably the most impressively impossible is the set of ideas known as quantum theory. This theory was developed in the early part of the twentieth century, because no one could find any other way to explain the behavior of atoms and molecules. One of the greatest physicists of recent times, Richard Feynman, described his feelings toward quantum theory in his book *QED*. "It is not a question of whether a theory is philosophically delightful, or easy to understand, or perfectly reasonable from the point of view of common sense," he wrote. The quantum theory "describes Nature as absurd from the point of view of common sense. And it agrees fully with experiment. So I hope you can accept Nature as She is—absurd. I'm going to have fun telling you about this absurdity, because I find it delightful."

As one example of the absurdity of quantum theory, consider a discovery made in 1970 by Sheldon Glashow, John Iliopoulos, and Luciano Maiani. Six years earlier, Murray Gell-Mann and George Zweig had proposed that the constituents of an atom's nucleus—the proton and the neutron—are composed of more fundamental particles, which Gell-Mann called "quarks." By 1970, the quark theory had become well known, but was not yet generally accepted.

Many properties of subatomic particles were well explained by the quark theory, but a few mysteries remained. One of these mysteries involved a particle called the neutral K-meson. This particle can be produced by particle accelerators, but it decays into other types of particles in less than a millionth of a second. The neutral K-meson was found to decay into many combinations of other particles, and everything that was seen made perfect sense in terms of the quark theory. The surprising feature, however, was something that was *not* seen: The neutral K-meson was never seen to decay into an electron and a positron. (A positron is a particle with the same mass as an electron, but with the opposite electrical charge—it is often called the "antiparticle" of the electron.) In the quark theory, this decay was expected, so its absence seemed to indicate that the theory was not working.

The quark theory held that there are three types of quarks, which were given the whimsical names *up, down,* and *strange.* (The word *quark* itself is associated with the number three; according to Gell-Mann, it was taken from the line, "Three quarks for Muster Mark!" in James Joyce's novel *Finnegans Wake.*) For each type of quark, there is also an antiquark. The neutral K-meson, according to the theory, is composed of a down quark and an anti-strange quark. The decay of a neutral K-meson into an electron and positron was expected to take place by a four-step process, as is illustrated below. There is no need to understand this process in detail, but for completeness, I have shown the individual steps. In addition to the quarks, the intermediate steps of the process involve particles called the neutrino, the W^+, and the W^-, but the properties of these particles will not be needed for what I want to say. The following diagram is to be read as a sequence of events, from top to bottom, starting with the quarks that make up the neutral K-meson:

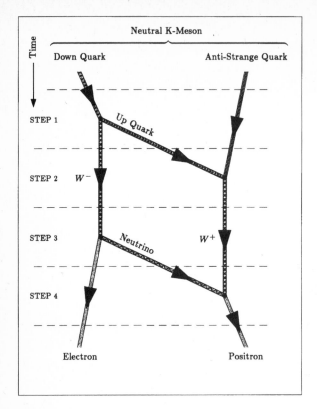

The Reaction That Doesn't Happen

In the first step, the down quark decays, or breaks up, into a W^- and an up quark. In the second step, the up quark combines with the anti-strange quark from the neutral K-meson to form a W^+ particle. The W^- particle decays in the third step into a neutrino and an electron, and in the fourth step, the neutrino combines with the W^+ to form a positron.

Puzzling over the diagram above, scientists could find nothing wrong with it. The quark content of the neutral K-meson was determined unambiguously by a variety of properties—it must be one down quark and one anti-strange quark. And all four steps in the process were thought to occur, although they had not been directly seen. In fact, the W^+ and W^- particles were not actually observed until 1983, when a mammoth experiment performed by a team of 135 physicists led to the observation of six W particles. Nonetheless, all four of the steps are also intermediate steps in other reactions—reactions that were known to happen. If any of the steps were impossible, then how could these other reactions take place? If the steps were possible, then what could stop them from taking place in the sequence shown above, producing a decay of a neutral K-meson into an electron and positron?

In 1970, Glashow, Iliopoulos, and Maiani proposed a solution to this puzzle. The solution is completely logical within the structure of quantum theory, yet it defies all common sense. It makes use of the strange way in which alternative processes are treated in quantum theory.

The physicists proposed that there is a fourth type of quark, in addition to the three types already contained in the theory. Such a fourth quark had already been suggested in 1964 by Glashow and James Bjorken, who were motivated by patterns in the table of known particles. The fourth quark had been called "charmed," a name that was revived in the much more specific proposal of Glashow, Iliopoulos, and Maiani. With the addition of another quark variety, the charmed quark, the neutral K-meson could decay into an electron and a positron by two distinct processes. The first would be the four-step process shown above; the second would be an alternative four-step process, in which the up quark produced in Step 1 and absorbed in Step 2 is replaced by a charmed quark.

Following the advice of the White Queen, it is now time to practice believing impossible things. The theory that includes the new quark allows two sequences of events, both beginning with a neutral K-meson and both ending with an electron and positron. The first sequence involves an up quark in Steps 1 and 2, and the second sequence involves a charmed quark in place of the up quark. According to the rules of common sense, the total probability for the decay of a neutral K-meson into an electron and positron would be the sum of the probabilities for each of the two sequences. If common sense ruled, the addition of the charmed quark would not help at all to explain why the decay is not seen. The rules of quantum theory, however, are very different from the rules of common sense.

According to quantum theory, if a specified ending can be achieved by two different sequences of events, then one calculates for each sequence a quantity called the "probability amplitude." The probability amplitude is connected to the concept of a probability, but the two have different mathematical forms. A probability is always a number between zero and one. A probability amplitude, on the other hand, is described by an arrow that one can imagine drawing on a piece of paper. The arrow is specified by giving its length and also its direction in the plane of the paper. The length must always lie between zero and one. If the specified ending can be achieved by only one sequence, then the probability is the square of the length of the probability amplitude arrow, and the direction of the arrow is irrelevant. For the decay of the neutral K-meson into an electron and positron, however, there are two sequences leading to the same result. In that case, the rules of quantum theory dictate that the tail of the second arrow is to be laid on top of the head of the first arrow, while both arrows are kept pointing in their original directions. A new arrow is then drawn from the tail of the first arrow to the head of the second, as shown:

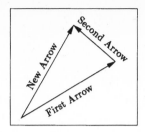

The total probability for the result is then the square of the length of the new arrow. Although this rule bears no resemblance to common sense, thousands of experiments have shown that it is indeed the way nature behaves.

For the decay of the neutral K-meson, Glashow, Iliopoulos, and Maiani proposed a definite procedure for calculating the way in which the charmed quark would interact with other particles. With this procedure, the probability amplitude arrow for the second sequence has the same length, but the opposite direction, as the arrow for the first sequence. When the two arrows are combined by the rules of quantum theory, the new arrow has zero length, corresponding to zero probability. Thus, by introducing an alternative mechanism through which the electron-positron decay could occur, it became possible to explain why the decay does not occur at all!

Although this explanation might not have been persuasive by itself, the decay discussed here was only one of about half a dozen processes that were expected but not observed. Glashow, Iliopoulos, and Maiani showed that the nonobservation of each of these processes could be explained by the charmed quark. The only drawback of the proposal was that none of the known particles appeared to contain a charmed quark. One must assume, therefore, that the charmed quark is much heavier than the other quarks, so that any particle containing a charmed quark would be too massive to have been produced in accelerator experiments.

In November 1974, a new particle with more than three times the mass of a proton was discovered simultaneously at the Brookhaven National Laboratory and the Stanford Linear Accelerator Center. The particle was called *J* on the East Coast and psi on the West Coast, so today it is known by the compromise name *J*/psi. The properties of this particle, by now demonstrated conclusively, show that it is composed of one charmed and one anti-charmed quark. The interaction properties of the charmed quark are exactly those predicted in 1970. Glashow, and the leaders of the two teams that discovered the *J*/psi, have all been awarded Nobel prizes in physics for their contributions. (Today, we believe that there are two more types of quarks, called "top" and "bottom," although the experimental evidence for top quark is not yet conclusive.)

The bizarre logic of quantum theory and the counterintuitive prediction of the charmed quark are only examples of the ideas that scientists are developing in their attempts to understand the world in which we live. The White Queen reigns throughout the world of science. The evidence so far indicates that nature obeys simple laws, but that these laws are very different from anything that one would be likely to imagine.

ALAN H. GUTH is a physicist and the Victor F. Weisskopf Professor of Physics at MIT, and a member of the National Academy of Sciences and the American Academy of Arts and Sciences. After receiving his Ph.D. in physics and doing nine years of postdoctoral research, Guth reached a turning point in his career when he invented a modification of the big bang theory called the inflationary universe. This theory not only explains many otherwise-mysterious features of the observed universe, but it also offers a possible explanation for the origin of essentially all the matter and energy in the universe. He has continued to work on the consequences of the inflationary theory, and has also explored such questions as whether the laws of physics allow the creation of a new universe in a hypothetical laboratory (probably yes, he thinks), and whether they allow the possibility of time travel (he would bet against it).

Symmetry: The Thread of Reality

Ian Stewart

~~~~~~~~ **S**ymmetry may seem to be just an unimportant repetition of structure, but its influence on the scientific vision of the universe is profound. Albert Einstein based all of his revolutionary theories of physics on the principle that the universe is symmetrical—that the laws of physics are the same at each point of space and each instant of time. Because the laws of physics describe how events occurring at one place and time influence events at other places and times, this simple requirement binds the universe together into a coherent whole. Paradoxically, as Einstein discovered, it implies that we cannot sensibly talk of absolute space and time. What is observed depends upon who observes it—in ways that are governed by those same underlying symmetry principles.

It is easy to describe particular kinds of symmetry—for example, an object has reflectional symmetry if it looks the same when viewed in a mirror, and it has rotational symmetry if it looks the same when rotated. Respective examples are the external shape of the human body, and the ripples that form on a pond when you throw a stone into it. But what is symmetry itself? The best answer that we yet have is a mathematical one: Symmetry is "invariance under transformations." A transformation is a method of changing something, a rule for moving it or otherwise altering its structure. Invariance is a simpler concept; it just means that the end result looks the same as the starting point.

Rotation through some chosen angle is a transformation, and so is reflection in some chosen mirror, so these special examples of symmetry fit neatly into the general formulation. A pattern of square tiles has yet another type of symmetry. If the pattern is moved sideways (a transformation) through a distance that is a whole-number multiple of the width of a tile, then the result looks the same (is invariant). In general, the range of possible symmetry transformations is enormous, and therefore, so is the range of possible symmetrical patterns.

Over the last one hundred and fifty years, mathematicians and physicists have invented a deep and powerful "calculus of symmetry." It is known as group theory, because it deals not just with single transformations, but with whole collections of them—"groups." By applying this theory, they have been able to prove striking general facts—for example, that there are precisely seventeen different symmetry types of wallpaper pattern (that is, repeating patterns that fill a plane), and precisely two hundred and thirty different types of crystal symmetry. And they have also begun to use group theory to understand how the symmetries of the universe affect how nature behaves.

Throughout the natural world, we see intriguing symmetrical patterns: the spiral sweep of a snail's shell; the neatly arranged petals of a flower; the gleaming crescent of a new moon. The same patterns occur in many different settings. The spiral form of a shell recurs in the whirlpool of a hurricane and the stately rotation of a galaxy; raindrops and stars are spherical; and hamsters, herons, horses, and humans are bilaterally symmetrical.

Symmetries arise on every conceivable scale, from the innermost components of the atom to the entire universe. The four fundamental forces of nature (gravity, electromagnetism, and the strong and weak nuclear forces) are now thought to be different aspects of a single unified force in a more symmetrical, high-energy universe. The "ripples at the edge of time"—irregularities in the cosmic background radiation—recently observed by the COBE (Cosmic Background Explorer) satellite help to explain how an initially symmetrical big bang can create the structured universe in which we now find ourselves.

Symmetrical structures on the microscopic level are implicated in living processes. Deep within each living cell there is a structure known as the centrosome, which plays an important role in organizing cell division. Inside the centrosome are two centrioles, positioned at right angles to each other. Each centriole is cylindrical, made from twenty-seven tiny tubes (microtubules) fused together along their lengths in threes, and arranged with perfect ninefold symmetry. The microtubules themselves also have an astonishing degree of symmetry; they are hollow tubes made from a perfect regular checkerboard pattern of units that contain two distinct proteins. So even at the heart of organic life we find the perfect mathematical regularities of symmetry.

There is another important aspect of symmetry. Symmetrical objects are made of innumerable copies of identical pieces, so symmetry is intimately bound up with replication.

Symmetries occur in the organic world because life is a self-replicating phenomenon. The symmetries of the inorganic world have a similarly "mass produced" origin. In particular, the laws of physics are the same in all places and at all times. Moreover, if you could instantly permute all the electrons in the universe—swapping all the electrons in your brain with randomly chosen electrons in a distant star, say—it would make no difference at all. All electrons are identical, so physics is symmetrical under the interchange of electrons. The same goes for all the other fundamental particles. It is not at all clear why we live in a mass-produced universe, but it is clear that we do, and that this produces an enormous number of potential symmetries. Perhaps, as Richard Feynman once suggested, all electrons are alike because all electrons are the self-same particle, whizzing backward and forward through time. (This strange idea came to him through his invention of "Feynman diagrams"—pictures of the motions of particles in space and time. Complex interactions of many electrons and their antiparticles often form a single zigzag curve in space-time, so they can be explained in terms of a single particle moving alternately forward and backward in time. When an electron moves backward in time, it turns into its antiparticle.) Or perhaps a version of the anthropic principle is in operation: Replicating creatures (especially creatures whose own internal organization requires stable patterns of behavior and structure) can arise only in mass-produced universes.

How do nature's symmetrical patterns arise? They can be explained as imperfect or incomplete traces of the symmetries of the laws of physics. Potentially, the universe has an enormous amount of symmetry—its laws are invariant under all motions of space and time and all interchanges of identical particles—but in practice, an effect known as "symmetry breaking" prevents the full range of symmetries from being realized simultaneously. For example, think of a crystal, made from a huge number of identical atoms. The laws of physics look the same if you swap the atoms around or move them through space and time. The most symmetrical configuration would be one for which all of the atoms are in the same place, but this is not physically realizable, because atoms cannot overlap. So some of the symmetry is "broken," or removed, by changing the configuration into one in which the atoms are displaced just enough to allow them to stay separate. The mathematical point is that the physically unrealizable state has a huge amount of symmetry, not all of which need be broken to separate out the atoms. So it is not suprising that some of that symmetry is still present in the state that actually occurs. This is where the symmetry of a crystal lattice comes from: the huge but unseen symmetries of the potential, broken by the requirements of the actual.

This insight has far-reaching consequences. It implies that when studying a scientific problem, we must consider not only what does happen, but what might have happened instead. It may seem perverse to increase the range of problems by thinking about things that don't happen, but situating the actual event inside a cloud of potential events has two advantages. First, we can then ask the question "Why does this particular behavior occur?"—because implicitly, this question also asks why the remaining possibilities did not, and that means we have to think about all the possibilities that don't occur as well as the ones that do. For instance, we can't explain why pigs don't have wings without implicitly thinking about what would happen if they did. Second, the set of potential events may possess extra structure—such as symmetry—that is not visible in the lone state that is actually observed. For example, we might ask why the surface of a pond is flat (in the absence of wind or currents). We will not find the answer by studying flat ponds alone. Instead, we must disturb the surface of the pond, exploring the space of all potential ponds, to see what drives the surface back to flatness. In that way, we will discover that nonflat surfaces have more energy, and that frictional forces slowly dissipate the excess, driving the pond back to its minimal-energy configuration, which is flat. As it happens, a flat surface has a lot of symmetry, and this, too, can best be explained by thinking about the "space" of all possible surfaces.

This, to me, is the deepest message of symmetry. Symmetry, by its very definition, is about what would happen to the universe if it were changed—transformed. Suppose every electron in your head were to be swapped with one in the burning core of the star Sirius. Suppose pigs had wings. Suppose the surfaces of ponds were shaped like Henry Moore sculptures. Nobody intends to perform actual experiments, but just thinking about the possibilities reveals fundamental aspects of the natural world. So the prosaic observation that there are patterns in the universe forces us to view reality as just one possible state of the universe from among an infinite range of potential states—a slender thread of the actual winding through the space of the potential.

IAN STEWART is one of the best-known mathematicians in the world. He is professor of mathematics at the University of Warwick, Coventry, where he works on nonlinear dynamics, chaos, and their applications, and has published over eighty research papers. He believes that mathematicians owe the public an explanation of what they are doing and why. He has written or coauthored over sixty books, including mathematics textbooks, research monographs, puzzle books for children, books on personal computers, and three mathematical comic books which have appeared only in French. Recent popular science books include *Does God Play Dice?; The Problems of Mathematics; Fearful Symmetry; Is God a Geometer?;* and *The Collapse of Chaos* (with biologist Jack Cohen). He writes the "Mathematical Recreations" column of *Scientific American;* is mathematics consultant to *New Scientist;* and contributes to magazines such as *Discover, New Scientist,* and *The Sciences.*

Professor Stewart also writes science fiction short stories for *Omni* and *Analog,* one of which was recently made into a play for Czech Radio.

Special Relativity: Why Can't You Go Faster Than Light?

W. Daniel Hillis

You've probably heard that nothing can go faster than the speed of light, but have you ever wondered how this rule gets enforced? What happens when you're cruising along in your spaceship and you go faster and faster and faster until you hit the light barrier? Do the dilithium crystals that power your engine suddenly melt down? Do you vanish from the known universe? Do you go backward in time? The correct answer is none of the above. Don't feel bad if you don't know it; no one in the world knew it until Albert Einstein worked it out.

The easiest way to understand Einstein's explanation is to understand the simple equation that you have probably seen before: $e = mc^2$. In order to understand this equation, let's consider a similar equation, one for converting between square inches and square feet. If i is the number of square inches and f is the number of square feet, then we can write the equation: $i = 144 f$. The 144 comes from squaring the number of inches per foot ($12^2 = 144$). Another way of writing the same equation would be $i = c^2 f$, where c in this case is equal to 12 inches per foot. Depending on what units we use, this equation can be used to convert any measure of area to any other measure of area; just the constant c will be different. For example, the same equation can be used for converting square yards to square meters, where c is 0.9144, the number of yards per meter. The c^2 is just the conversion constant.

The reason why these area equations work is that square feet and square inches are different ways of measuring the same thing, namely area. What Einstein realized, to everyone's surprise, was that energy and mass are also just two different ways of measuring the same thing. It turns out that just a little bit of mass is equal to a whole lot of energy, so in the equation, the conversion constant is very large. For example, if we measure mass in kilograms and energy in joules, the equation can be written like this: $e = 90{,}000{,}000{,}000{,}000{,}000\ m$. This means, for example, that a charged-up battery (which contains about one million joules of energy) weighs about 0.0000000001 grams more than a battery that has been discharged.

If we work with different units, the conversion constant will be different. For instance, if we measure mass in tons, and energy in BTUs, then c will be 93,856,000,000,000,000. (It happens to work out that the conversion constant in a particular set of units is always the speed of light in those units, but that is another story.) If we measure both energy and mass in what physicists call "the natural units" (in which $c = 1$), we would write the equation: $e = m$, which makes it easier to understand; it just means that energy and mass are the same thing.

It doesn't matter whether the energy is electrical energy,

chemical energy, or even atomic energy. It all weighs the same amount per unit of energy. In fact, the equation even works with something physicists called "kinetic" energy, that is, the energy something has when it is moving. For example, when I throw a baseball, I put energy into the baseball by pushing it with my arm. According to Einstein's equation, the baseball actually gets heavier when I throw it. (A physicist might get picky here and distinguish between something getting heavier and something gaining mass, but I'm not going to try. The point is that the ball becomes harder to throw.) The faster I throw the baseball, the heavier it gets. Using Einstein's equation, $e = mc^2$, I calculate that if I could throw a baseball one hundred miles an hour (which I can't, but a good pitcher can), then the baseball actually gets heavier by 0.000000000002 grams—which is not much.

Now, let's go back to your starship. Let's assume that your engines are powered by tapping into some external energy source, so you don't have to worry about carrying fuel. As you get going faster and faster in your starship, you are putting more and more energy into the ship by speeding it up, so the ship keeps getting heavier. (Again, I should really be saying "massier" not "heavier" since there is no gravity in space.) By the time you reach 90 percent of the speed of light, the ship has so much energy in it that it actually has about twice the mass as the ship has at rest. It gets harder and harder to propel with the engines, because it's so heavy. As you get closer to the speed of light, you begin to get diminishing returns—the more energy the ship has, the heavier it gets, so the more energy that must be put into it to speed it up just a little bit, the heavier it gets, and so on.

The effect is even worse than you might think because of what is going on inside the ship. After all, everything inside the ship, including you, is speeding up, getting more and more energy, and getting heavier and heavier. In fact, you and all the machines on the ship are getting pretty sluggish. Your watch, for instance, which used to weigh about half an ounce, now weighs about forty tons. And the spring inside your watch really hasn't gotten any stronger, so the watch has slowed way down so that it only ticks once an hour. Not only has your watch slowed down, but the biological clock inside your head has also slowed down. You don't notice this because your neurons are getting heavier, and your thoughts are slowed down by exactly the same amount as the watch. As far as you are concerned, your watch is just ticking along at the same rate as before. (Physicists call this "relativistic time contraction.") The other thing that is slowed down is all of the machinery that is powering your engines (the dilithium crystals are getting heavier and slower, too). So your ship is getting heavier, your engines are getting sluggish, and the closer you get to the speed of light, the worse it gets. It just gets harder and harder and harder, and no matter how hard you try, you just can't quite get over the light barrier. And that's why you can't go faster than the speed of light.

W. DANIEL HILLIS was cofounder and chief scientist of Thinking Machines Corporation, and the principal architect of the company's major product, the Connection Machine. At Thinking Machines, his research focused on parallel programming, applications of parallel computers, and computer architecture. Hillis's current research is on evolution and parallel-learning algorithms. For example, he has used simulations that are models of biological evolution to automatically produce efficient sorting programs from random sequences of instructions. His long-term interests are focused on designing a thinking machine, and biological simulation provides a new approach to the problem.

Hillis is a member of the Science Board of the Santa Fe Institute, the Advisory Board of the *OS Journal on Computing,* and the External Advisory Board of the Institute for Biospheric Studies. He is a fellow of the Association of Computing Machinery, and a fellow of the American Academy of Arts and Sciences. He is the holder of thirty-four U.S. patents and is an editor of several scientific journals, including *Artificial Life; Complexity; Complex Systems;* and *Future Generation Computer Systems.* He is author of *The Connection Machine.*

Part Six

The Future

How Long Will the Human Species Last?
An Argument with Robert Malthus and Richard Gott
Freeman Dyson

$\mathbf{A}$lmost two hundred years ago, the Reverend Thomas Robert Malthus published his famous "Essay on the Principle of Population as It Affects the Future Improvement of Society." Malthus took a gloomy view of the human situation. He started by stating two general laws. First, he said, population always tends to increase geometrically, that is to say, the increase each year is a fixed fraction of the population. Second, he said, the amount of food available grows arithmetically, which is to say, the food supply each year increases by a fixed amount independent of the population. It is, then, mathematically certain that the geometrical increase of population will overtake the arithmetical increase of food. Malthus deduced from his two laws the prediction that population growth will in the end be held in check only by famine, war, and disease. The less fortunate members of our species will always be living in poverty on the brink of starvation.

Malthus was an English clergyman, one of the founding fathers of the science of economics. His gloomy predictions caused economics to be known as "the gloomy science." So far as England was concerned, his predictions turned out to be wrong. After his essay was published, the population of England grew fast, but the supply of food and other necessities grew faster. It remains to be seen whether his predictions will prove to be correct for humankind as a whole. The weak point of his argument was his failure to distinguish between a simple mathematical model and the complicated real world.

It is now two hundred years later, and my friend Richard Gott has published another gloomy essay. It appeared in May 1993 in the British scientific journal *Nature,* with the title "Implications of the Copernican Principle for Our Future Prospects." Richard Gott is a famous astronomer who is, like me, interested in questions concerning the existence and destiny of life in the universe. By the "Copernican Principle," he means the principle that human beings do not occupy a special position in the universe. The principle is named after Copernicus, the astronomer who first put the sun rather than the earth at the center of the solar system. After Copernicus's heretical theory had become orthodox, we gradually learned that we are living on an ordinary planet orbiting around an ordinary star situated in an outlying region of an ordinary galaxy. Our situation is in no way special or central. The Copernican Principle, according to Richard Gott, asserts that there is nothing extraordinary about our situation either in space or in time.

To apply the Copernican Principle to the human situation in time, Gott relies on the mathematical notion of "a priori probability." The a priori probability of an event is the probability that we would calculate for it to occur if we were totally ignorant of any special circumstances or causes that might influence it. The Copernican Principle states that the place we are living in, and the time we are living at, should not have small a priori probability. For example, a central position has a much smaller a priori probability than a noncentral position in a galaxy. Therefore, the Copernican Principle says that we should be living in a noncentral position, as in fact we are. When the Copernican Principle is applied to our situation in time, it says that Richard Gott and I should not be living either close to the beginning or close to the end of the history of the human species. We should be living roughly halfway through human history. Since we know that the species originated about two hundred thousand years ago, our living at a nonspecial time implies that the future duration of our species should also be of the order of two hundred thousand years. This is a very short time compared with the future lifetime of the earth or the sun or the galaxy. That is why I call Richard's argument gloomy. He argues that humans are not here to stay, that we are likely to be extinct in a few million years at most, and that we are not likely to survive long enough to colonize and spread out over the galaxy.

Gott's argument has the same weak point as Malthus's argument. Both of them use an abstract mathematical model to describe the real world. When Gott applies his Copernican Principle to the actual situation in which we find ourselves, he is assuming that a priori probabilities can be used to estimate actual probabilities. In reality, we are not totally ignorant of the special circumstances surrounding our existence. Therefore, the a priori probabilities may give poor estimates of what is actually likely or unlikely to happen.

After I read Richard Gott's essay and thought about it for a while, I wrote him the following letter. The letter explains why I disagree with his conclusions, although I find his argument important and illuminating.

Dear Richard,

Good morning, Mister Malthus. Thank you for sending me your delightfully gloomy essay on the implications of the Copernican Principle. The logic of your argument is impeccable. I have to admire the way you deduce such a rich harvest of conclusions from such a simple and apparently plausible hypothesis. As you say, the hypothesis is testable and, therefore, a legitimate scientific hypothesis. Your reverend predecessor Thomas Robert Malthus likewise deduced dire consequences from a testable hypothesis. My judgment is that in both cases, the hypothesis is a little too simple to be true. Nevertheless, in both cases, the study of the hypothesis's consequences is useful, since the consequences stand as a warning to an improvident species.

I remember as a child in the 1930s, when Britannia still ruled the waves, being puzzled by the paradox of the improbability of my personal situation. I was born into the ruling class of the ruling empire, which meant that I was born into the top one tenth of a percent of the world's population. The probability that I should have been so lucky was only one in a thousand. How could I explain my luck? I never found a satisfactory explanation. The notion that I should have had an equal a priori probability of being born anywhere in the world came into conflict with the fact that I was born an upper-class Englishman. In this case, it seemed that the notion of equal a priori probabilities did not work.

I think the paradox of my birth arises also in the hypothesis of your essay. The notion of equal a priori probability only makes sense when we are totally ignorant of particular facts that may bias the probabilities. If we happen to know that a particular improbable event has in fact happened, then the probabilities of all related events may be drastically altered.

Let me give you an artificial example. Suppose we discover two planets inhabited by alien civilizations in remote places in the sky. We send out an embassy to make contact and establish a human settlement on one of the two planets. Because we find it too expensive to send embassies to both, we choose the planet to be visited by a toss of a coin. Now, it happens that the inhabitants of Planet A are ogres who immediately devour all human visitors, while the inhabitants of Planet B are benevolent zoologists who place the visitors in a zoo and encourage them to breed, and shield them from all possible harm for a billion years. These facts about the aliens are unknown to us when we make the choice of which planet to visit. Now, according to your Copernican hypothesis, the probability that we extend the life of our species by visiting Planet B is vastly smaller than the probability that we visit Planet A and become extinct within a few million years through the normal vicissitudes of life on Planet Earth. But we know that in this case, the probabilities of A and B are equal. The point of the example is that the knowledge of a single improbable fact, in this case the existence of Planet B with its benevolent inhabitants, makes the Copernican estimate of a priori probability of lifetimes completely wrong.

To return from this artificial example to our actual situation, it seems to me that the Copernican hypothesis is already falsified by our knowledge of one improbable fact, the fact that we happen to live at the precise moment, within a few hundred years, when the technical possibility arose for life to escape from a planet and spread over the universe. Whether the human species happens to take advantage of this possibility is here irrelevant. The point is that the possibility exists now and did not exist for the previous four billion years. We are, as it happens, born at a lucky time, whether we take advantage of it or not. The knowledge of this improbable fact changes all the a priori probabilities, because the escape of life from a planet changes the rules of the game life has to play.

Another independent falsification of the Copernican hypothesis arises from the fact that we happen to be living at the birth of genetic engineering. In the past, as Darwin remarked, the majority of creatures left no surviving progeny because species did not interbreed. Extinction of species implied the extinction of progeny. In the future, the rules of the game may be different. Genetic engineering may blur and erode the barriers between species, so that extinction of species no longer has a clear meaning. Copernican assumptions about the longevity of species and of progeny are then no longer valid.

In conclusion, let me say that I am not contesting the logic of your argument, but only the plausibility of your assumptions. If the Copernican hypothesis is true, then the consequences that you deduce from it follow. I am delighted that you have stated the consequences in such a clear and forthright manner. I am only anxious that the public should not be led to believe that your hypothesis is an established fact. The example of your predecessor Malthus shows that this may be a real danger. Uncritical belief in Malthus's predictions helped to hold back political and social progress in England for a hundred years. I trust you not to allow a similar uncritical belief in your conclusions to hold back progress today.

Thank you for giving me this opportunity to think about important questions in a new way.

Yours sincerely,

Freeman Dyson

This letter will not be the last word in the argument between me and Richard. I am now looking forward to Richard's reply.

FREEMAN DYSON is a professional scientist who has made a second career writing books for nonscientists. He likes to vary his scientific activities with forays into engineering, politics, arms control, history, and literature. He writes mostly about people and things he has seen, sometimes about people and things he has imagined. He was born and raised in England, worked during the Second World War as a boffin for Royal Air Force Bomber Command, emigrated after the war to America, and became famous by solving some esoteric mathematical problems in the theory of atoms and radiation.

Since 1953, he has been professor of physics at the Institute for Advanced Study in Princeton, New Jersey. The main theme of his life is the pursuit of variety. This includes variety of people, variety of scientific theories, variety of technical tricks, variety of cultures, and languages. He is author of five books addressed to the general public: *Disturbing the Universe; Weapons and Hope; Origins of Life; Infinite in All Directions;* and *From Eros to Gaia.*

The Uniqueness of Present Human Population Growth

Joel E. Cohen

T wo numbers summarize the present size and growth of Earth's human population. The total number of people on Earth is approximately 5.5 billion. The increase in the number of people on Earth from last year to this is around 92 million. Using these two numbers alone, I show you that the present growth of population could not have occurred over long periods of the past and cannot continue long. The human species is passing through a brief, transient peak of global population growth that has no precedent and probably will be unique in all of human history.

To derive such a grand conclusion requires only elementary arithmetic, the two numbers (5.5 billion people, an annual increase of some 92 million) and—of course—one or two innocent assumptions. The arithmetic involved will be familiar if you understand compound interest.

I begin with the past. To extrapolate to the past without any additional data requires some assumption about how the population changed in earlier years. As always in making assumptions, various choices are possible.

One possible assumption is that the recent increase in population of 92 million also occurred in each prior year. If this assumption were true, the population two years ago would be this year's population minus twice 92 million, or (5.5 − 2 × 0.092) billion = 5.3 billion. Given this assumption, Earth's population must have been nearly zero between 59 and 60 years ago, because 59.1 years × 92 million per year = 5,496 million or very nearly 5.5 billion people. This is an absurd conclusion. Earth's population was nowhere near zero around 1930, and surely the 1930s were no Garden of Eden. If the population had been near zero in 1930, it could not possibly have increased by 92 million in one year. Adam and Eve were fertile but not, according to the best available account, *that* fertile. Conclusion: The absolute change in numbers, 92 million, from last year to this is much greater than the absolute increases that must have occurred over most of human history.

A second possible assumption is that the same *relative* growth rate of population from last year to this occurred in each prior year. By definition, this year's relative growth rate is the change in population from last year to this, divided by last year's population. So the increase of 92 million divided by last year's population of roughly 5.4 billion works out to a relative growth rate of 1.7 percent per year. If the relative growth rate of population had been the same throughout the past, then Adam and Eve must have lived less than 1,300 years ago, because a population starting from two people and increasing by 1.7 percent per year for 1,290 years would grow to more than 5.5 billion people.

Even Archbishop James Ussher, primate of all Ireland, who calculated that God created the universe in 4004 B.C., would agree that human history is much longer than 1,290 years and, therefore, would have had to concede that the present relative growth rate could not have been sustained for most of human history. (Ussher determined the moment of creation to be October 23 at noon. The Oxford divine and Hebraist John Lightfoot revised Ussher's calculations and concluded that creation occurred on October 26 at 9 A.M., the date and time which are quoted in many texts. I wish to avoid taking sides in the controversy over whether the creation occurred October 23 at noon or October 26 at 9 A.M.)

Firm conclusion: The current increase of Earth's human population, in either absolute or relative terms, vastly exceeds the average increases that occurred over most of human history.

As a further confirmation that the present relative growth rate of population could not have been sustained for most of human history, let's play just one more arithmetic game before we turn to the future. The last ice age ended roughly 12,000 years before the present (actually, at different times in different places). Archaeological evidence confirms that there were substantial human populations on all the continents except Antarctica. If there had been only one person 12,000 years ago, and that population had grown (parthenogenetically, at first) at an average rate of 1.7 percent per year, then Earth's current population would be approximately 7,100,000,000,000,000,000,000,000,000,000,000,000, 000,000,000,000,000,000,000,000,000,000,000, 000,000,000,000,000,000, or 7.1×10^{87} people. This is more than 10^{78} times as many people as Earth currently has. If the average weight per person were 50 kilograms (about 110 pounds), the mass of people would be about 6×10^{64} times the mass of Earth (which is a mere 6×10^{24} kilograms), and Earth would have left its current orbit long ago.

I must confess that no one knows either the exact present population of Earth or its exact rate of increase. How sensitive are these conclusions to errors in the two numbers on which the calculations are based? Would it make much difference if the actual population of Earth this year were 5.0 billion or 6.0 billion? Would it make much difference if the relative growth rate were 1.4 percent per year or 2.0 percent per year? (I know of no professional demographer who suggests that the current population size or growth rate lie outside these ranges.) The answer, in every case, is the same. Even if the current data are not exactly correct, the human population's current absolute growth rate (in numbers per year) and the current relative growth rate (in percent per year) could not have been sustained over most of history.

Human demographic history differed from the hypothetical models of constant absolute growth or constant relative growth in at least two important ways. First, population growth differed from one place to another. While Babylonian and Hittite cities were rising in what are now called Iraq and Anatolia (Asian Turkey), what is now Europe probably saw no comparable demographic growth, and parts of South America may have been still unpeopled.

Second, population growth varied over time: now faster, now slower, sometimes even negative. It has been conjectured that the first real surge in human population growth occurred before 100,000 B.C., when people discovered how to use and make tools. Archaeological and historical evidence shows that a surge occurred from 8000 B.C. to 4000 B.C., when people discovered or invented agriculture and cities, and again in the eighteenth century, when people discovered science and industry and major food crops were exchanged between the New World and the Old. Between these periods of rapid rise were much longer periods of very slow growth or occasional falls (as in the fourteenth century, when the Black Death struck).

By far the largest surge in human history started shortly after World War II and still continues. According to the best estimates, the human population increased from somewhere between two million and twenty million 12,000 years ago to about 5.5 billion today. The relative growth rate of global population accelerated so strikingly in recent centuries that roughly 90 percent of the increase in human numbers during the last twelve millennia occurred since A.D. 1650, in less than 350 years. Until as recently as 1965, human numbers have grown like the money in a bank account with an erratically rising interest rate.

Now let's look into the future. I confidently assert that the average growth rate of the human population for the next four centuries cannot equal its present growth rate. Why? Because if the present growth rate of 1.7 percent per year persists for four hundred years (plus or minus ten), the population will increase at least a thousandfold, from 5.5 billion now to 5.5 trillion people.

The total surface area of Earth, including oceans, lakes, streams, icecaps, swamps, volcanoes, forests, highways, reservoirs, and football fields, is 510 million square kilometers. With a population of 5.5 trillion, each person would be allotted a square area less than 10 meters on a side. This area is perhaps commodious as a jail cell, but it is incapable of supporting a person with the food, water, clothing, fuel, and physical and psychological amenities that distinguish people from ants or bacteria. No optimist, if that is the right word, has suggested that Earth could support 5.5 trillion people.

The present growth rate of 1.7 percent per year is an average over some quickly growing regions (like Africa and southern Asia) and some slowly growing regions (like Europe, Japan, and northern America). In detailed projections published in 1992, the United Nations considered what would happen if each region of the world maintained its 1990 levels of fertility and gradually reduced its death rate. Naturally, the faster-growing regions will grow faster than the slowly growing regions. Therefore, the faster-growing regions will become a larger and larger fraction of the global population pie, and the average growth rate of the global population will increase. At first, the U.N.'s hypothetical population doubles within forty years. Within 150 years, it passes 600 billion, and by 2150, it exceeds 694 billion people.

The U.N. commented dryly: "To many, these data would show very clearly that it is impossible for world fertility levels to remain at current levels for a long time in the future, particularly under assumptions of continuing mortality improvement."

Many people, I among them, think that upper limits like 600 billion people, let alone 5.5 trillion, far exceed what humans and Earth would ever tolerate. I give you simple arguments with very large limits not to suggest that the limits I mentioned are anywhere near the actual limits, but to illustrate that even with extremely large limits, the amount of time remaining to the human population to bring its numerical (not spiritual, cultural, or economic) growth to a halt is not extremely long. In the next few to tens of decades, a drastic though not necessarily abrupt decline in the global population growth rate is inevitable.

The global population growth rate can fall from its present value of 1.7 percent per year to zero or below only by some combination of a decline in birth rates and a rise in death rates in those largely poor areas of the world with currently high fertility. (Forget extraterrestrial emigration. To achieve a reduction in the present global population growth rate from 1.7 percent to 1.6 percent would require the departure of 0.001 $\times$ 5.5 billion or 5.5 million astronauts in the first year and more every year after that. The cost of exporting that many people would bankrupt the remaining Earthlings and would still leave a population that doubled every 46 years. Demographically speaking, space is not the place.)

People face a choice: lower birth rates or higher death rates. Which would you prefer?

JOEL E. COHEN is professor of populations at The Rockefeller University, New York. His research deals mainly with the demography, ecology, population genetics, and epidemiology of human and nonhuman populations and with mathematics useful in those fields. He is author of *A Model of Simple Competition; Casual Groups of Monkeys and Men: Stochastic Models of Elemental Social Systems; Food Webs and Niche Space; Community Food Webs: Data and Theory;* and a book of scientific jokes, *Absolute Zero Gravity* (with Betsy Devine).

This essay is based on a book entitled *How Many People Can the Earth Support?,* to be published in 1995 by W. W. Norton and Company, New York.

Who Inherits the Earth?
An Open Letter to My Sons
Niles Eldredge

Dear Doug and Greg,

Every generation thinks the world is going to hell in a handbasket—an obscure saying, but you know what it means: Things ain't what they used to be, as Duke Ellington so succinctly put it. And now that we approach a new millennium, we are all thinking, "Maybe this time the news really *is* bad. Maybe the world really is going to hell in a handbasket."

You have grown up going out in the field with me collecting 380-million-year-old trilobites, brachiopods, and many other long-extinct vestiges of truly ancient life. You know that extinction and evolution go hand in hand. And you have also seen the awesome exuberance of life. You know that, hard as the knocks that living systems have taken over the eons have been, life is terrifically resilient. It always bounces back—at least so far, and we are talking about a 3.5-*billion*-year track record.

And you also know of, and have become personally concerned about, the rising wave of extinction that has begun to engulf the millions of species living right now. Life may be resilient, and may make it through this latest bout of mass extinction, eventually bursting forth in a fresh evolutionary riot of form and color. But that is scant comfort to us in the here and now—especially to you who still have the bulk of your lives ahead of you.

You know the statistics nearly as well as I do. Species are disappearing at the rate of twenty-seven thousand a year (three every hour!), according to E. O. Wilson at Harvard. Looking back over the past 540 million years, we now realize that all previous episodes of mass extinction came through habitat disruption and ecosystem collapse. Before humans appeared on the scene, abrupt climatic change (with a great burst of cometary impact thrown in on at least one occasion) caused mass extinction events. Today, we see with equal clarity, it is our own species, *Homo sapiens,* that is the real culprit.

We are cutting down, burning off, plowing under, and paving over the surface of the land at an expanding rate. We are transforming terrestrial ecosystems into agricultural monocultures. We are creating vast concrete, steel, plastic, and glass environments devoid of life—save that of our fellow humans and a few commensal species that seem to thrive on the periphery of our existence.

We are diverting streams, and our agricultural and industrial runoff is poisoning our rivers, lakes, and now, even our oceans. And you know, too, of the direct and dangerous side effects that our industrial activities are having on the atmosphere: global warming from "greenhouse gases" (like carbon dioxide) and ozone holes caused by chemical reaction between ozone (O^3) and chlorofluorocarbons—such as the freon we rely on so heavily to cool our houses, offices, and cars during the increasingly hot summer months.

Direct destruction of habitats through human action is an exact parallel of climate change altering ecosystems of the past—alterations that triggered the relatively sudden disappearance of vast numbers of species. We are locked into a vicious circle; we have a seemingly insatiable collective appetite and the apparent endless ability to exploit and "develop" resources with ever-increasing efficiency. And every time we make a breakthrough, our population soars.

That's a vicious circle. And you can see that the real cause of such rampant human destruction of the natural world is our out-of-control population growth. Ten thousand years ago, at the dawn of agriculture, there were no more than one million people on Earth. There are now 5.7 billion—and the number is zooming. More people require ever more resources; development of more resources—with a rare but important exception that I'll get to in a moment—begets more people.

The problem, of course, is that our own species is at risk. If the population explosion doesn't do in the civilized world—through famine, warfare, and even disease (despite the miracles of modern medicine)—our own species still faces the very real possibility of going down along with millions of other species in a mass extinction of our very own devising. For it is a mistake to assume, as so many of us still do, that we are no longer a part of the natural world—and that whatever happens to all those ecosystems and species out there is of no consequence to ourselves.

That's what we have been telling ourselves for the past ten thousand years or so—ever since agriculture was first invented in the Middle East and settled existence based on a predictable food supply became possible, bringing with it a panoply of human achievements—among them, the invention of the written word. We are fortunate to have the

Judaeo-Christian Bible, one of but a few documents with roots in this vital transitional period, for it was the invention of agriculture that forever changed the human stance toward nature. And the folks who wrote what we call "the Bible" knew that and wrote about it.

Genesis has many stories, including two and a half versions of how the cosmos came into being. It tells us that we were created in God's image and were ordained to have "dominion" over the earth and all its creatures. But I am convinced we have created God in our image rather than the other way around. I need not remind you that I am convinced the physical earth has had a very long history—a sort of "evolution" all its own, in concert with the development of the solar system, our Milky Way galaxy, and the universe itself. You also know of my conviction that all species— living and extinct—have descended from a single common ancestor that arose well over 3.5 billion years ago by a natural process of biological evolution. You also know my thoughts on *Homo sapiens:* we have evolved, like any other species.

We don't read Genesis today for an accurate glimpse of the history of the cosmos and earthly life. But we should read it carefully for its views about who we were in the minds of some of our ancestral savants all those millennia ago. For there is an essential truth in Genesis—a recognition that humans had altered their place in the natural world.

All species, with the sole exception of our own, are broken up into relatively small populations; each of these populations is integrated into a local, dynamic ecosystem. The squirrels in Central Park are more concerned with the health of the oak trees and whereabouts of barn owls in their neighborhood than with the well-being of members of their own species living across the river in New Jersey. Our own ancestors were no different, and still today there are some peoples (themselves on the brink of imminent extinction) who live in local populations and play definitive roles in local ecosystems. The Yanomani—currently being slaughtered by gold miners in Amazonian Venezuela and Brazil—are a case in point.

The fact is that agriculture changed all that. With agriculture we essentially and effectively declared war on local ecosystems. All plants save the one or two species we were cultivating all of a sudden became "weeds." All animals save the few we domesticated and those we still occasionally hunted became "pests." We were, it seemed, liberated from the natural world. We were no longer dependent upon its yearly bounty. We could afford, we felt, to disregard it, to feel that we had escaped it, championed over it. We had achieved "dominion."

As this shift occurred, population began to rise; 5.7 billion from 1 million in a scant ten thousand years is amazing! And, for quite a while, we seemed to be getting away with our newfound freedom. We could trash environments with seeming impunity. We would abandon settlements as the soil became exhausted, taking up residence elsewhere. We were really letting nature repair the damage—ample sign (one would think) that we were not as independent of nature as we liked to believe.

But we are now, if not exactly wall-to-wall people, rapidly approaching the point where there are too few resources to support the global human population. Thomas Malthus saw this coming way back at the end of the eighteenth century. Some economists, like Julian Simon of the University of Maryland, persist in denying that we are in trouble at all. After all, they claim, the wealth from industrialization tends to stabilize population growth. But there is no hope that the living standards of the Third World will ever be brought up to our own (declining) level. And as it is, we in the privileged wealthy nations are consuming something like thirty times the resources per capita than someone living, say, in Bangladesh. In a very real sense, the United States population should be multiplied by thirty to measure the true impact that its size has on the world economy.

All of which sounds dismal. It sounds like I really do think the world is going to hell in a handbasket. But I don't think so, or, I should say, I don't *necessarily* think so. Here's why.

Agriculture was only a dramatic step in a long chain of cultural evolutionary events in human prehistory. Ever since the advent of material culture some 2.5 million years ago, human ecological history can be read as one long, and often very clever and very successful, attempt to "do something about the weather." The advent of fire, for example, enabled our ancestors *Homo erectus* to leave Africa just under a million years ago—traveling north into the very teeth of a major glacial advance to hunt the large, newly evolved beasts of Ice Age Europe.

Agriculture was not a plot against nature—it was just a continuation of a culturally adaptive strategy to render food resources more reliable. The point is that the myth—the story of who we are and how we stand toward nature—that we read in Genesis is an explanation of the newly achieved status quo. Everybody, all the time, needs such stories, such explanations, simply to function.

The Genesis myth was a good story for its time. I actually think the concept of God as it has come down to us from those distant times reflects a need to invent something all-encompassing and powerful to make up for our consciously stepping out of the confines of the local ecosystem. Humans still living within ecosystems are prone to invest spirits in the species around them—but otherwise have no need for a comforting or frightening Almighty God.

But that is an aside. Otherwise, the myth of dominion seemed to fit the facts pretty well. It really did look like we had escaped nature—to the point where we could deny that we were ever a part of it in the first place.

But the story won't work anymore. And therein lies the real hope for the future; we need to adjust that story. We need to see that we never really left nature but just redefined our role in it. We see, all too clearly, that we can no longer trash local ecosystems with impunity. Even *The New York Times* sees that, when they acknowledged the other day that draining and pollution of wetlands starves and poisons the coastal seafood so vital to our economy and sustenance.

We are unique in being a global species that interacts with the totality of the global environment. Internally, we exchange one trillion dollars every day on a global basis. And that one trillion resonates against the natural world.

Meanwhile, the global system—the atmosphere, hydrosphere, lithosphere, and biosphere—struggles to maintain the status quo (not consciously, of course—but through the natural interactive cycles that fall out of plain physics and chemistry). The health of the global ecosystem is nothing more—nor less—than the collective health of all the component local ecosystems, all linked together through a complex web of energy flow. Cut down the rain forest and you alter rainfall patterns and the flow of nutrients to the sea. Poison the oceanic surface and you eliminate the major source of daily replenishment of atmospheric oxygen. It seems almost laughable that we could ever think we had escaped nature. But we see it now: We have radically altered our stance toward, and our position within, nature, but we never did manage to escape it. And now we are threatening ourselves, and a host of our fellow species, with extinction.

What to do? Stabilize population. How? Economic development tends to do that. But it is already too late to think realistically in such terms; there is no way to raise Third World economies up to present industrialized-nation economic status. But there are encouraging signs that *education* has the same effect. Especially the education, enfranchisement, and economic empowerment of women. And the data increasingly show that women enthusiastically will abandon multiple births if given alternative pathways to living.

It is especially apt that education is the key. We have come to our present pass honestly—and actually through a great deal of cleverness. Alone among species (I firmly believe), we have cognition and true culture. We have used our culture, our cleverness, our smarts, to keep playing around with and improving our ecological adaptation—our way of making a living, of dealing with the often harsh realities of physical existence.

We should just go on with the theme of trying to do something about the weather. But we have to adjust our myth about who we are and how we fit into the cosmos. We must acknowledge that we are a part of nature—and that we occupy a unique position as a global species. We have to break the natural biological cycle that always sees population size increase when access to resources is itself increased. We have to wise up—see ourselves for who we are—*Homo sapiens,* "wise mankind."

If we acknowledge that we don't own the earth, if we moderate ourselves, restore ecosystems, and let other species live, there is still a good chance that we—along with our fellow species—can survive to inherit the earth. It's a big challenge, but a winnable one. There are some hopeful signs. Ozone damage is already reversing, for example, because people have awakened and taken concerted, definitive action. It will be up to your generation to complete the switch to an altered, more accurate vision of who we are and how we fit into the natural world.

Good luck, guys!

Dad

NILES ELDREDGE has been an active research paleontologist on the staff of the American Museum of Natural History since 1969. He has devoted his entire career to effecting a better fit between evolutionary theory and the fossil record. In 1972, he and Stephen Jay Gould announced the theory of punctuated equilibria. Since then, Eldredge has developed his views on the hierarchical structure of living systems, and the nature of the relation between ecology and evolution. Most recently, he has focused on the mass extinctions of the geological past, and their implications for understanding the modern biodiversity crisis.

Eldredge has published over one hundred and sixty scientific articles, reviews, and books, notably *The Monkey Business* on scientific creationism; *Time Frames*, his account of punctuated equilibria; *Fossils*, in which he provides a scientific autobiography for a general audience; *The Miner's Canary*, which explores the relation between mass extinctions of the geological past and the present-day, human-engendered biodiversity crisis; and *Interactions*, with philosopher Marjorie Grene), an attack on the ultra-Darwinian underpinnings of sociobiology. His most recent book is *Evolutionary Confrontations*, an examination of current controversies in evolutionary biology.

Eldredge regularly lectures at universities and research centers on aspects of evolutionary theory and biodiversity issues, and for the American Museum travel program. He is an active bird-watcher and an avid trumpet and cornet player and collector.

Can Science Answer Every Question?

Martin Rees

C an science, eventually, answer every question? Among those who thought not was the French philosopher Auguste Comte. More than one hundred years ago, he gave an example of an unanswerable question as: "What are the stars made of?" And he was quickly proved wrong. Even before the nineteenth century was over, astronomers had realized how to find the answer. When starlight is passed through a prism and spread into a spectrum, we see the telltale colors of different substances—oxygen, sodium, carbon, and the rest. Stars are made of the same kinds of atoms that we find on Earth. Arthur C. Clarke once said, "If an elderly scientist claims something is impossible, he is almost surely wrong." Comte was just one of them.

Ninety-two different kinds of atoms occur on Earth, but some are vastly more common than others. For every ten atoms of carbon, you'd find, on average, twenty of oxygen, and about five each of nitrogen and iron. But gold is a million times rarer than oxygen, and others—platinum and mercury, for instance—are rarer still. Remarkably, these proportions are roughly the same in the stars.

So where did the different kinds of atoms come from? Is there a reason why some are commoner than others? The answer lies in the stars themselves, whose centers are hot enough to fulfill the alchemist's dream—to transmute the base metals into gold.

Everything that has ever been written in our language is made from an alphabet of just twenty-six letters. Likewise, atoms can be combined in a huge number of different ways into molecules: some as simple as water (H_2O) or carbon dioxide (CO_2), others containing thousands of atoms. Chemistry is the branch of science that studies just how this happens.

The most important ingredients of living things (ourselves included) are carbon and oxygen atoms, linked (along with others) into long chainlike molecules of huge complexity. We couldn't exist if these particular atoms weren't common on Earth.

Atoms are themselves made up of simpler particles. Each kind has a specific number of protons (with positive electric charge) in its nucleus, and an equal number of electrons (with negative electric charge) orbiting around it; this is called the atomic number. Hydrogen is number 1; uranium is number 92.

Since all atoms are made up of the same elementary particles, it wouldn't be surprising if they could be changed into one another. This can occur, for instance in a nuclear explosion, but atoms are so "robust" that they aren't themselves destroyed by the chemical changes that occur in living things, or in scientists' laboratories.

No natural process on Earth can create or destroy the atoms themselves. The basic building blocks, the chemical elements, have the same proportions as when the solar system formed about four and a half billion years ago. We'd like to understand why the atoms were "dealt out" in these particular proportions. We could leave it at that—perhaps the creator turned ninety-two different knobs. But scientists always seek simple explanations and try to trace complex structures back to simple beginnings. In this instance, astronomers have supplied the key insights. It seems that the universe indeed started out with simple atoms, which were fused and transmuted into the heavier ones inside stars.

The sun and other stars are giant spheres of gas. Inside them, two forces are competing against each other: gravity and pressure. Gravity tries to pull everything toward the center, but when gas is squeezed, it heats up, and the pressure gets so high that it can balance gravity. To supply enough pressure, the sun's center must be far hotter than the surface we see—about fifteen million degrees, in fact.

Heat leaking from its hot center keeps the sun shining; the "fuel" is the same process that makes hydrogen bombs explode. Hydrogen, the simplest element, has a nucleus of just one proton. In a gas that's as hot as the sun's core, individual protons crash together so hard that they stick. This process converts hydrogen into helium (atomic number 2). The energy release in a star is steady and "controlled," not explosive as in a bomb. This is because gravity pulls down the overlying layers firmly enough to effectively "hold the lid on," despite the huge pressure in the stellar core.

Fusion of hydrogen into helium releases so much heat that less than half the sun's central hydrogen has been used up, even though it's been shining for four and a half billion years. But stars heavier than the sun shine much more brightly. Their central hydrogen gets used up (and turned into helium) more quickly: in less than a hundred million years. Gravity then squeezes these stars further, and their central temperature rises still higher, until helium atoms can themselves stick together to make the nuclei of heavier atoms— carbon (six protons), oxygen (eight protons), and iron (twenty-six protons). An old star develops a kind of onion-skin structure, where the inner hotter layers have been transmuted into the heavier nuclei.

This, anyway, is what astronomers calculate should happen to a star. But you might wonder how these theories can be tested. Stars live so long compared with astronomers that we see just a single "snapshot" of each one. But we can test our theories by looking at the whole population of stars. Trees can live for hundreds of years. But if you'd never seen a tree before, it wouldn't take you longer than an afternoon wandering around in a forest to deduce the life cycle of trees. You would see saplings, fully grown specimens, and some that had died. Astronomers use just that kind of reasoning to check their ideas about how stars evolve. We can observe clusters of stars that all formed at the same time but are of different size. We can also observe gas clouds where even now new stars, perhaps with new solar systems, are forming.

But not everything in the cosmos happens slowly. When its fuel is used up, a big star faces a crisis—its center collapses, triggering a colossal explosion that blows off the outer layers, with speeds of ten thousand kilometers per second, making a supernova.

When a nearby star explodes as a supernova, it flares up for a few weeks to be far brighter than anything else in the night sky. The most famous such event was observed in A.D. 1054. In July of that year, the Chinese chief astronomer, Yang Wei-Te, addressed his emperor in these words: "Prostrating myself before Your Majesty, I have observed the appearance of a guest star. On the star there was a slightly iridescent yellow color." Within a month, the "guest star" started to fade. At that place in the sky, there now lies the Crab Nebula—the expanding debris from this explosion. This nebula will remain visible for a few thousand years; it will then become too diffuse to be seen, and merge with the very dilute gas and dust that pervade interstellar space.

These events fascinate astronomers. But why should anyone else care about explosions thousands of light-years away? It turns out that if it wasn't for supernovae, the complexities of life on planet Earth couldn't exist—and we certainly wouldn't be here.

The outer layers of a star, when a supernova explosion blows them off, contain all the atoms that can be made, starting just from hydrogen, by the processes that kept it shining over its entire lifetime. What's gratifying is that, according to the calculations, this mixture should contain a lot of oxygen and carbon, plus traces of many other elements. The predicted "mix" is remarkably close to the proportions now observed in our solar system.

Our galaxy, the Milky Way, is a huge disk one hundred thousand light-years across and containing a hundred billion stars. The oldest stars in it formed about ten billion years ago from the simple atoms that emerged from the big bang—no carbon, no oxygen. Chemistry would then have been a very dull subject. Our sun is a middle-aged star. Before it even formed, four and a half billion years ago, several generations of heavy stars could have been through their entire life cycles. The chemically interesting atoms—those essential for complexity and life—were forged inside these stars. Their death throes, supernova explosions, flung these atoms back into interstellar space.

After wandering for hundreds of millions of years, these

"debris" atoms from early supernovae might have joined a dense interstellar cloud, which collapsed under its own gravity to form stars, some surrounded by retinues of planets. One such star would have been our sun. Some atoms might have found themselves in the newly forming earth, where they could have been recycled through many forms of life. Some may now be in human cells—including yours. Every carbon atom—those in every cell of your blood, or in the ink on this page—has a pedigree as old as the galaxy.

A galaxy resembles a vast ecological system. Pristine hydrogen is transmuted, inside stars, into the basic building blocks of life—carbon, oxygen, iron, and the rest. Some of this material returns to interstellar space, thereafter to be recycled into new generations of stars.

Why are carbon and oxygen atoms so common here on Earth, but gold and uranium so rare? This everyday question isn't unanswerable—but the answer involves ancient stars that exploded in our Milky Way more than five billion years ago, before our solar system formed. The cosmos is a unity. To understand ourselves, we must understand the stars. We are stardust—the ashes from long-dead stars.

MARTIN REES is an astrophysicist and cosmologist, and research professor in the United Kingdom's Royal Society. He was a professor at Cambridge University for nineteen years, and director of the Cambridge Institute of Astronomy for ten years. He has also worked at Sussex University in England, and in the United States, has held visiting positions at Harvard, Caltech, the Princeton Institute for Advanced Study, and the University of California. He belongs to a number of foreign academies and is actively involved in projects to promote international collaboration in science.

Professor Rees has remained at the forefront of cosmological debates. He has had several important ideas on how stars and galaxies form, how to find black holes, and on the nature of the early universe. He is now trying to understand the mysterious "dark matter" which seems to fill intergalactic space—it is the gravitational pull of this dark matter which will determine whether our universe expands forever or eventually collapses to a "big crunch." He has always been interested in the broader philosophical aspects of cosmology. For instance: Why does our universe have the special features that allowed life to evolve? Are there other universes, perhaps governed by quite different physical laws? He has written and lectured widely on these subjects, for both specialist and general audiences. He discusses them more fully in a forthcoming book. He is also the author, with John Gribbin, of *Cosmic Coincidences: Dark Matter, Mankind, and Anthropic Cosmology.*

About the Editors

JOHN BROCKMAN, president of Edge Foundation and founder of The Reality Club, is a writer and literary agent. His works include *By the Late John Brockman; 37; Afterwords;* and *The Third Culture*. He is editor of *About Bateson* and four volumes in the Reality Club book series: *Speculations; Doing Science; Ways of Knowing;* and *Creativity*. He is publisher and editor of *Edge,* the newsletter of the Edge Foundation.

KATINKA MATSON, a writer and literary agent, has written three books: *The Working Actor; Short Lives: Artists in Pursuit of Death;* and *The Psychology Today Omnibook of Personal Development*.